WINNING WITH WATER

WINNING WITH WATER

Soil-Moisture Monitoring For Efficient Irrigation

by
Gail Richardson, Ph.D.
and
Peter Mueller-Beilschmidt, P.E.

An INFORM Report

INFORM, Inc.
381 Park Avenue South
New York, NY 10016
(212) 689-4040

Copyright © 1988 by INFORM, Inc.
All rights reserved

Library of Congress Cataloging-in-Publication Data

Richardson, Gail.
    Winning With Water.

    1. Irrigation efficiency. 2. Soil moisture—
Measurement. 3. Irrigation efficiency—West (U.S.)
4. Soil moisture—West (U.S.)—Measurement.
I. Mueller-Beilschmidt, Peter. II. Title.
S619.E34R53   1988   631.7   88-8846
ISBN 0-918780-42-X (pbk.:alk.paper)

INFORM, Inc., founded in 1973, is a nonprofit research organization that identifies and reports on practical actions for the protection and conservation of natural resources and public health. INFORM's research is published in books, abstracts, newsletters and articles. Its work is supported by contributions from individuals and corporations and by grants from over 40 foundations.

Printed on recycled paper
Cover design by Saul Lambert
Book design by James Carr
Photographs by Gail Richardson
Drawings by Ron Kopels
Graphs by Katharine Grant and John Mensing

Contents

Preface	ix
Acknowledgements	xi

Part I. Introduction and Findings

	1. *A Method for Seeing Beneath the Soil*	3
	2. *The Rewards of Seeing Beneath the Soil*	9

Part II. Background

	3. *Western Irrigation and Its Problems*	19
	4. *What the Field Surface Doesn't Show*	27

Part III. Tools and Methods

	5. *Gypsum Blocks*	37
	6. *Installing Gypsum Blocks on Farm Fields*	47
	7. *Gathering and Graphing Gypsum-Block Data*	58

Part IV. INFORM's Field Work

	8. *Cooperators and Fields*	63
	9. *Using Gypsum-Block Data to Improve Water Distribution*	70
	10. *Using Gypsum-Block Data to Schedule Irrigations and Reduce Water Use*	84

Part V. Benefits Achieved

	11. *Water Reductions and Yield Increases*	97
	12. *Benefits and Costs of the Soil-Moisture Method*	103

Appendices

A.	*Detailed Results of INFORM's Demonstrations on 32 Fields*	113
B.	*Two Additional Examples of Uneven Distribution*	154
C.	*Why 10 Fields Were Not Tested*	158
D.	*Flow Rates Through Siphons*	160
E.	*How INFORM's Cooperators Evaluated the Soil-Moisture Method*	162
F.	*Letter From the Westlands Water District Evaluating Gypsum Blocks After Comparing Them with a Neutron Probe in 1986 Field Tests*	166
G.	*Report on 1987 Field Demonstrations of the Soil-Moisture Method Sponsored by the Yolo County Resource Conservation District*	168

Tables

1. Four Tools for Monitoring Soil Moisture — 45
2. Crops Studied by INFORM on 32 California Fields — 66
3. Major Soil Types Found on INFORM's Fields and Their Average Water-Holding Capacities — 68
4. Water Reductions and Crop Yields in 21 Tests — 100
5. Water-Cost Savings on INFORM's Field Strips Where Yields Were Higher or the Same — 104
6. Value of Measured Yield Increases on INFORM's Field Strips — 106
7. Combined Water-Cost Savings and Values of Increased Yields on INFORM's Field Strips — 107
8. Comparison of Per-Season Costs of INFORM's Test Method and Farmers' Streamlined Method — 109

Graphs

1. How to Interpret Gypsum-Block Readings — 38
2. Moisture Changes Between Readings — 60
3. An Evenly Watered Alfalfa-Field Strip — 72
4. Sulfurized Irrigations Improve Water Penetration on an Alfalfa-Field Strip — 74
5. An Impermeable Layer on an Alfalfa-Field Strip — 77
6. Using Slower Irrigations to Wet the Center Section of a Tomato Bed — 80
7. Improved Distribution Reduces Irrigation Frequency — 82
8. Gypsum-Block Data Show that Smaller Siphons Safely Reduce Water Use on a Tomato-Field Strip — 86
9. Monitoring the 3-Foot Depth on an Alfalfa-Field Strip to Apply One Instead of Two Irrigations per Growth Cycle — 88
10. Monitoring the 2-Foot Depth on a Cotton-Field Strip to Eliminate Three Irrigations — 90
11. Bracketing Procedures on an Alfalfa-Field Strip — 92
12. An Underirrigated Tomato-Field Strip — 155
13. Overwatering Obscures Patterns of Water Distribution on an Alfalfa-Field Strip — 156

Preface

In a single century, dams and deep turbine pumps have brought irrigation to 50 million arid and semi-arid acres in 17 western states. Irrigation on this scale creates some of the West's greatest economic and ecological challenges, and farmers are under many pressures to do the best possible job of managing their water supplies.

Most farmers admit that overwatering of crops does occur and adds to costly problems of yield damage, drainage buildup, salinity and pollution of groundwater and surface streams. But they also know that the hardest part of finding field-level solutions is discovering which questions to ask.

To begin with, each field's soil types and conditions are unique and pose different puzzles of irrigation management. In addition, weather and water-supply conditions change each year. Finally, everything farmers do to fields, from plowing to harvesting, affects, and is affected by, their irrigation practices.

The method of soil-moisture monitoring with gypsum blocks that INFORM has demonstrated since 1984 gives farmers a reliable way to find the right questions to

Preface

ask about their irrigation practices on each field. It shows them where they underwater or overwater field sections so they can test and improve their irrigation practices to reduce these problems. It enables them to track water use by crops between irrigations and plan their irrigation schedules efficiently.

By observing changes in soil-moisture levels, farmers can learn where their roots are deep and active or where they are shallow and cramped. This improves their ability to tend to the needs of the hidden as well as the visible half of their crops.

This soil-moisture method is inexpensive, easy to learn and flexible. Farmers can adapt it at their own pace to their own needs. They need not measure their water applications directly, but can readily work to reduce overwatering simply by observing and changing the rate, duration and number of irrigations they apply.

The method does not add a new set of tasks to a farmer's already long list of field responsibilities. Rather it gives farmers a systematic way to use more efficiently their pumps, pipelines, hydrants, ditches, siphons, fertilizers, pesticides, labor and machinery to achieve the overall goal of farming—high yields.

Finally, the soil-moisture method tested by INFORM is designed especially for use on fields irrigated by flooding all or parts of field surfaces. These fields constitute about three-quarters of the West's irrigated acreage. If farmers managing these fields adopted the soil-moisture method of irrigation management, then soils and crops, farmers and non-farmers alike would greatly benefit.

It is our hope that the findings on the benefits of the soil-moisture method, contained in this report, will play an important role in advancing the use of the method in the years ahead.

Joanna Underwood
Executive Director

Acknowledgements

The INFORM-tested method of improving irrigation practices originated from a question posed in 1977 by California farmer Frank Sieferman. Sieferman, then Director and now President of the Yolo County Resource Conservation District, asked Peter Mueller-Beilschmidt, an irrigation engineer: How can farmers find out for themselves if they are overwatering crops, and how can they get field proof of the benefits of water conservation?

Peter's answer involved using gypsum blocks to monitor soil-moisture changes, as described in this report, and his demonstrations on several Yolo County fields in 1978-79 indicated the method's practicality. Then INFORM undertook to confirm and document the method's benefits on many fields and soil types and to ensure its broad availability to western farmers.

INFORM learned of Peter Mueller-Beilschmidt's work in 1983, thanks to Kevin Wolfe of Friends of the River; Bob McBride, an independent environmental researcher; and Nick Arguimbau, Esq., then of Citizens for a Better Environment. Nick and Bob helped INFORM devise a research plan using the soil-moisture method,

Acknowledgements

and in 1984 INFORM asked Peter Mueller-Beilschmidt to conduct pilot tests of his method on four fields near Davis. The promising results of this work, reported in *Saving Water from the Ground Up* (INFORM, 1985) led INFORM to expand the study to 18 fields in 1985 and 1986.

INFORM's field research and findings depended on the help of and evaluations by the following farmers: Tony Barcello, Bryan Barrios, Leroy Bertolero, Harry Dewey, Myron Fagundes, Craig Fulwyler, Randy Gafner, Bill Longfellow, Delbert Mello, Bill and Chester Roth, Ron Timothy, Gary Wilson, and Wayne Wisecarver.

Several other individuals not only followed INFORM's research with interest but have also added to the available information about gypsum blocks and soil-moisture monitoring, with different kinds of support from INFORM.

• Byron Steinert and Don Upton of the Westlands Water District arranged for field comparisons of gypsum-block readings with high-tech neutron-probe readings, to evaluate the uniformity of the soil-moisture monitoring tool used in INFORM's tests.

• Gerald Robb and Tracy Slavin, the water conservation and management staff of the Westlands Water District, conducted these tests in 1986 and expect to publish a more extensive analysis of the results summarized in Appendix F.

• John Tiedeman and Walter Cheechov of the U.S. Soil Conservation Service (SCS) demonstrated the soil-moisture method to farmers on 14 fields in the Sacramento Valley in 1987, under the auspices of the Yolo County Resource Conservation District (RCD), as reported in Appendix G.

Acknowledgements

- William Beatty, Area Conservationist of the SCS Area IV in California, comprising most of the Sacramento Valley, received requests from 14 RCDs in his area for field demonstrations with gypsum blocks in 1988 as a result of his interest in the soil-moisture method.

- Arturo Carvajal of California's Mobile Agricultural Water Conservation Laboratory, and Randy Gafner, President of the Pond-Shafter-Wasco Resource Conservation District of Kern County, used gypsum blocks in 1987 to evaluate water-distribution problems on 12 fields that were also being evaluated by the Mobile Lab. The Pacific Gas and Electric Company provided funding for this work.

- Wes Robbins, head of the SCS's Ogallala Water-Management Team for Colorado, helped in 1987 to expand farmers' knowledge of the INFORM-tested method on the eastern plains of Colorado, where he has also developed his own methods of teaching farmers to use gypsum blocks to conserve an endangered aquifer and reduce their pumping costs.

Numerous other individuals have also given INFORM valuable advice and encouragement along the way. These include Orville L. Abbott, Executive Officer and Chief Engineer of the California Water Commission; Jean Auer of the Commonwealth Club of California; Charles M. Benbrook, Executive Director of the National Board on Agriculture; Vashek Cervinka, Research Manager of the California Department of Food and Agriculture; Daniel M. Dooley, former Chairman of the California Water Commission; Harrison C. Dunning, Professor of Law at the University of California, Davis; Roy Foote, retired District Conservationist, and Robert Fry, Resource Conservationist, both of the

Acknowledgements

SCS in Kings County; Stan Gale of Del Rio Farms; Thomas Graff, Esq., of the Environmental Defense Fund; Betty Harris, Executive Director of the California Association of Resource Conservation Districts (CARCD); D.W. Henderson, a retired professor from the Department of Land, Air and Water Resources of the University of California, Davis; Earl Hess, Water Management Engineer of the SCS in Colorado; Richard E. Howitt, Professor of Agricultural Economics at the University of California, Davis; Laura King of the Natural Resources Defense Council; Phillip LeVeen, Lecturer in Agricultural Economics, University of California, Berkeley; John Merriam, retired professor from California Polytechnic State University at San Luis Obispo and one of the West's foremost authorities on surface irrigation; Chris Nelson of the San Felipe Ranch; Roy Rutz, former President of CARCD; Mary Shallenberger of the California Senate Natural Resources and Wildlife Subcommittee; Clarence Tighe of the Elkhorn Ranch; and Henry J. Vaux, Jr., Professor of Resource Economics at the University of California, Riverside.

Ralph E. Grossi, President of the American Farmland Trust; Dr. I. Garth Youngberg, Executive Director of the Institute for Alternative Agriculture, Inc.; and Edwin H. Clark II, of the Conservation Foundation, have helped us understand the larger context of our work.

The debts to INFORM's California field staff are substantial. Nancy Lowe Barker of Hanford collected field data for INFORM in 1985 and processed data in 1986 as well as supervising field staff. Jonathan Barker put in long hours creating computer programs for data storage and analysis. Nancy and Jonathan were also the gracious hosts for INFORM's educational field days in 1985 and 1986. Richard Molina and Ray Diehl gave INFORM invaluable, high-quality field help in 1986.

Acknowledgements

INFORM's New York staff provided essential support throughout the project and the writing of this report. Joanna Underwood, INFORM's Executive Director, worked tirelessly to ensure the project's success, and never allowed complex mechanics to cloud a broader vision. Nancy Lilienthal, INFORM's Research Director, brought good humor, care and intelligence to her reviews and corrections of the manuscript. Mary Redway gave valuable assistance as Coordinator of INFORM's Western Irrigation Project. Peter Green edited the final report and made many useful suggestions. Perrin Stryker mastered the technical details himself in order to convey the essence of our work to readers of *INFORM Reports*. James Carr designed the report, and organized the production in cooperation with Arthur Hamparian. The staff of the Fund for the City of New York and its Nonprofit Computer Exchange was always helpful and cooperative during the production process. Ron Kopels rendered rough sketches into finished drawings. Charles Lowy and John Mensing cheerfully typed a small library of proposals and manuscripts. Rob Young, Blair Simpson, Larry Naviasky, Ellen Poteet, Rebecca Cooney and Katharine Grant were on hand when it counted.

INFORM's Directors knew when to slow us down to keep us on course and when to push us along to meet our goals. Special thanks go to Anthony Wolff, for his good counsel, and to Jay Last and Martin Krasney, California residents, who provided many other kinds of help as well.

Without the generous support of a number of foundations and individuals, however, INFORM's work in this area would not have been possible. We owe particular thanks to: Dr. George M. Byrne; Compton Foundation, Inc.; Michael J. Connell Foundation; The Wallace Alexander Gerbode Foundation; John A. Harris

Acknowledgements

IV; Melvin B. Lane, Publisher; The Max and Anna Levinson Foundation; Jessie Smith Noyes Foundation; The David and Lucile Packard Foundation; The San Francisco Foundation; Gordon L. Smith Trust; The Tides Foundation; and Wallace Genetic Foundation, Inc.

Part I
Introduction and Findings

1
A Method for Seeing Beneath the Soil

Western farmers irrigate 45 to 50 million acres using an estimated 85% of the annual water supply of 17 western states. Yet these farmers cannot see what happens to their water after it soaks into fields and thus cannot easily tell if they are overwatering crops.

Winning with Water describes INFORM's field tests of a method farmers can use to "see" into the soil to observe and correct inefficient practices and thereby reduce the guesswork and waste in irrigation. The method was developed by Peter Mueller-Beilschmidt for use on crops irrigated by flooding field surfaces or portions of them. Surface-irrigated fields comprise three-quarters of the West's irrigated acreage.

The INFORM-tested method uses small, low-cost electronic tools called gypsum blocks to monitor increases in soil-moisture levels caused by irrigations and decreases caused by crop water consumption. Each gypsum block is a plaster-of-paris plug containing two electrodes. When a block is buried in a crop-root zone, it loses and absorbs moisture similarly to the surrounding soil. Two insulated wires that are connected to the block's electrodes are extended to the surface.

By attaching the projecting wires to a pocket-size, battery-powered impedance meter, the *relative* moisture changes in the block and the surrounding soil are tested: More current passes through the completed circuit from the meter to the block when the soil and block are wet.

Farmers who follow INFORM's procedures use gypsum blocks to manage irrigations by installing them at several locations and depths in a field's crop-root zone. They read the blocks weekly or twice weekly to see whether they are irrigating unevenly and/or too frequently. Then, by experimenting over one or two seasons with practices intended to correct these problems, farmers learn which adjustments reduce their water use, cut energy and water costs and improve yields. They conduct these tests on small field sections so they can observe the impact of changes on yields before deciding which practices should be extended to entire fields.

Once a farmer develops an efficient set of practices for a field, these can be maintained season after season using far fewer monitoring locations than during the farmer's initial experiments. Then the costs for gypsum blocks and labor run about $2 per acre or less and the benefits can easily amount to $50 or more per acre in lower water and energy costs and improved yields.

Sometimes the gains from using the method can be spectacular: An INFORM test on one strip of a 160-acre alfalfa field indicated that by using about 30% less water on the entire field the farmer could have earned $20,000 or more in higher hay yields alone.

Although gypsum blocks have been developed and improved over several decades, their financial benefits for farmers have never been adequately demonstrated to encourage their broad use on western fields. Such wide-

A Method for Seeing Beneath the Soil

spread application could not only improve farmers' income but help them to reduce the expensive and sometimes dangerous problems linked with irrigation:

• Falling groundwater tables under 16 million farm acres

• Salt damage to crops on one-third of the West's acres

• The buildup of salty farm drainage beneath hundreds of thousands of farm acres

• The deterioration of drinking water supplies due to the leaching of salts, pesticides and natural toxic elements from farm soils into groundwater and surface streams.

Alfalfa's remarkable roots, with top parts shown, can extend 10 to 15 feet into good, deep soils to extract water. Yet they normally draw more than half of the crop's supply from soil lying within 3 feet of the surface.

INFORM's 1984-1986 Field Demonstrations

Between 1984 and 1986, with Peter Mueller-Beilschmidt as technical consultant, INFORM demonstrated the soil-moisture method on 32 commercial fields belonging to 16 farmers in California's Central Valley. All but one field were surface irrigated. They included a wide variety of soil types, soil conditions and irrigation problems.

INFORM concentrated its study on alfalfa, cotton and tomatoes because these are among California's most important crops. They are grown on about one-third of the state's 9.9 million irrigated farm acres. They use about one-third of the state's annual water supply, an amount sufficient to meet the annual drinking and sanitation needs of about 65 million urban dwellers.

INFORM accepted farmers as cooperators in the study on a first-come basis. Sources of recommendation included the U.S. Soil Conservation Service and/or neigh-

Part I: Introduction and Findings

Delbert Mello, a Fresno County farmer (left), learned from Peter Mueller-Beilschmidt that one section of his salt-blighted alfalfa field is overwatered and another section is underwatered. Soil-moisture data from buried gypsum blocks allow farmers to "see" the effects of their irrigation practices on crop root zones.

boring farmers. INFORM's cooperators ranged in age from 21 to 80 years and owned or managed farms comprising 500 to over 2,000 acres. Of the 12 farmers who worked most actively with INFORM, and accounted for 27 of the 32 fields, one participated in the study for three growing seasons, seven for two seasons and four for one season.

INFORM chose California for the field demonstrations because it irrigates more acres, uses more irrigation water and confronts more severe irrigation problems than any other western state. And it has the highest production of farm products, by number and value, of any state in the U.S.

The Two-Strip Method for Testing Irrigation Improvements

On each field tested for irrigation improvements, INFORM's basic procedures included:

• Selecting two equal-size strips, each comprising less than 1 to more than 12 acres and stretching from the water source to the drain ditch. One strip was identified as "INFORM's" and the other as "owner's" without regard to their possible differences in soil type or other field conditions.

• Installing gypsum blocks at three locations on each strip and, at each location, implanting two sets of blocks at 1-foot depths down to four feet

• Gathering soil-moisture readings twice weekly from each strip

• Using the soil-moisture data from both strips to analyze patterns of water distribution, and using the data

A Method for Seeing Beneath the Soil

from the INFORM strip to test water reductions on that strip

• Comparing water use, energy use and yields on the less-watered INFORM strip with results on the owner's strip being managed by the farmer using his standard practices.

Most of the fields lying along the Kings River in California's San Joaquin Valley are irrigated by flooding soil surfaces. The soil-moisture method tested by INFORM was developed for use on surface-irrigated fields, which comprise three-quarters of the West's 50 million irrigated acres.

Water-Distribution Analyses

Before testing water reductions, INFORM analyzed water-distribution patterns on each of the 32 fields and found that only seven fields were being evenly watered on both strips. The remaining 25 fields had problems of uneven water distribution that showed up clearly in soil-moisture data as partial dryness in some locations and depths in the crop-root zone on one or both strips.

Part I: Introduction and Findings

As INFORM's data indicate, such problems are common on surface-irrigated fields. They often occur because farmers apply water at volumes and rates poorly matched to their soil types. They may also have more complex causes such as impenetrable soil layers on some field sections, soil that has been compacted by heavy farm machines, or salt.

On unevenly watered fields, where crop dryness is apparent first on the sections that are partially skipped by irrigations, farmers often apply larger and more frequent irrigations than would be needed if distribution patterns were more uniform. Hence, solving distribution problems is an essential first step toward improving irrigation efficiency.

Water-Reduction Tests

INFORM tested fewer and/or lighter irrigations on test strips in 22 of the 32 fields. (See Appendix C for reasons why 10 fields studied were not tested.) One of the 22 strips was tested twice, making a total of 23 season-long experiments.

Two of the 23 tests were invalidated for different reasons. (See page 97.) In all, 21 valid water-reduction tests were performed.

INFORM estimated water use on test strips and owners' strips from information about outflow rates of the siphons or gated pipelines used for irrigating most of the fields. However, no practical method was available for measuring outflow rates through the hydrants used on one-fourth of INFORM's test fields.

INFORM analyzed the yields of alfalfa strips by counting bales after monthly hay cuttings. Yields on row-crop fields, chiefly cotton and tomatoes, were measured or estimated by the farmers themselves.

2
The Rewards of Seeing Beneath the Soil

*I*NFORM's 1984-86 irrigation tests showed the effectiveness of the soil-moisture method for improving farmers' irrigation practices.

Water Reductions and Crop Yields

INFORM's striking results on 21 test strips, compared to the results of farmers' standard practices on owners' strips on the same fields included the following:

• Nineteen tests reduced water applications by 6 to 58%. In 10 tests, water use was reduced by 20 to 40%. (On two fields belonging to the same farmer, INFORM's strips used fewer irrigations than the owner's. But no estimates of percentage reductions were possible because the farmer's irrigations varied greatly in application rate and duration.)

• Ten test strips produced higher yields and six produced equal yields. (Four of the remaining five tests resulted in lower yields due to poor soil conditions on the test strip and/or to communications and logistical problems. The fifth test's yield results were discounted by the farmer, who said that the soil types on the tested and monitored strips were too different to make valid comparisons.)

• Seven alfalfa strips showed higher hay production with one watering per monthly growth cycle than that achieved with two waterings per month on the owners' strips.

Benefits

Dollar benefits produced by reducing water and increasing yields, in the 14 field tests where one or both benefits were achieved and could be measured on INFORM's strip, included:

• From under $1 to over $165 per acre in reduced water costs and/or higher yields for one season

• From $1 to $39 per acre for water-cost savings on the 12 field strips where water volumes were estimated. On the one strip tested twice, savings over two seasons rose to $90 per acre.

• From $5.60 to $124 per acre for increased hay production on seven alfalfa-field strips; and $126.50 per acre for increased cotton production on one cotton-field strip.

(See page 103 for reasons why benefits were not estimated on seven of the 21 tested strips.)

INFORM estimated labor-cost reductions of $1.50 to $3 per acre for each "saved" irrigation. Yet, the complexities of labor use on irrigated fields precluded detailed analysis of these savings. (See pages 107-108.)

INFORM's cooperators reported many unmeasured (or unmeasurable) benefits on all 32 fields studied. These included less crop damage due to overwatering, higher-quality tomato yields due to less bloating, better information about soil types and better means of assessing and using more efficiently different kinds of irrigation equipment and soil-treatment methods.

Costs

INFORM's seasonal costs for blocks, at $5 apiece, and labor time for installing and reading the blocks, at $5 an hour, amounted to $4 to $12 per acre on the 27 to 160-acre fields tested. The auger and hammer set used for installing the blocks cost $175 and the meter cost $200.

However, INFORM learned from one cooperator, Ron Timothy of Solano County, that these costs dropped steeply over three seasons after he adapted the method to his own needs. Timothy's modifications included using fewer blocks on each field strip; using data from some fields to manage others having similar soil types; and using fewer, longer and slower irrigations.

Timothy now spends about $1.25 per acre per growing season for blocks and labor, and another 25¢ per acre for equipment repairs, to manage irrigations on 600 acres of processing tomatoes. (He ignores the costs of his one-time purchase of the auger, hammer and meter because these costs have fallen to a few cents per acre after being used on hundreds of acres over several seasons.) Timothy's yearly after-cost benefits from lower pumping costs and payroll are $30.50 per acre and $18,300 for all 600 acres.

Since 1984, Ron Timothy of Solano County has adapted the soil-moisture method to improve his irrigation efficiency and yields on 600 acres of processing tomatoes. He uses fewer blocks and monitoring sites than in INFORM's tests and nets $18,300 per season in reduced labor and pumping costs alone.

Part I: Introduction and Findings

Untested Opportunities for Water and Cost Savings

On every test strip, INFORM found opportunities for further reducing water use and water costs that could have been explored during a second growing season of trial-and-error tests. (Only one field was tested twice. See Appendix A, Field 24.) For example, on all field strips, a "bracketing" procedure could have been used to pin down the minimum water application needed with each irrigation. (See page 85.) Eight to 10 strips of row crops could have used fewer irrigations if early-season irrigations had been spaced farther apart. And lighter "pre-irrigations" could have sufficiently wetted six cotton fields before planting.

In general, it takes more than one season to find the most efficient combination of number, rate, volume and length of irrigations. Each change in one of these elements requires one irrigation for evaluating results (using moisture data) and deciding whether other changes are needed. Thus, whenever INFORM's strip skipped one of the five to seven irrigations that farmers typically applied during the test period, an opportunity for experimenting with other kinds of changes was "lost."

Farmers' Reactions

Farmers themselves gave INFORM proof of the practical appeal of the soil-moisture method for surface irrigators by developing their own uses for the method as INFORM's tests proceeded:

• Ron Timothy used blocks to eliminate an unneeded irrigation of sugar beets that had "overwintered" in the field, before harvesting them in June 1985. He saved $2,000 in unspent pumping costs on an 80-acre field.

- Bill Longfellow of Kings County used blocks to schedule two fewer waterings on a 10-acre cotton field during the 1986 season while INFORM was testing two of his other fields. His experimental field produced "very good" yields.

- Randy Gafner, who manages Oran Gil Farms in Kern County, used block data to learn that sulfurized water applications opened up severely clogged sections of the root zone on the salt-blighted alfalfa field being tested by INFORM.

- Craig Fulwyler, who manages McConnell Farms in Kern County, developed a special cotton-harvesting plan to discover that yields on INFORM's strip, watered five times during the season, were worth $126.50 more per acre than yields from cotton growing only three to five rows distant and watered seven times. (See Appendix A, Field 16.)

- Gary Wilson of Wilson Ag in Kern County used block data to learn that his new high-tech surge-valve pipelines reduced water applications on the owner's strip of an INFORM-tested cotton field, but did not significantly improve irrigation uniformity compared to standard gated pipelines.

Siphons and bales helped INFORM show farmers how to grow hay by using less water. Siphons of different diameters draw water at different rates from supply ditches onto fields. Small siphons that apply water slowly often reduce waste. Bales counted before their removal from fields allowed INFORM to compare yields on less-watered and more heavily watered field strips.

The Next Steps

The INFORM-tested, gypsum-block method has proved itself reliable and beneficial in field applications. Broad educational efforts could now give tens of thousands of western farmers an opportunity to benefit from it.

Part I: Introduction and Findings

Many federal, state, and local agencies and private companies are already experienced in water, soil and/or energy conservation on farms. If the staffs of these agencies trained themselves in the method, they could readily educate large numbers of farmers. These agencies and companies include the following:

• County-level, farmer-run soil and water conservation districts that are for most farmers the primary source of information about, and training in, conservation methods

• The U.S. Soil Conservation Service whose technical staff are expert advisors to soil and water conservation districts

• The Federal-State Cooperative Extension, whose county-level farm educators regularly run classes and workshops for farmers on a variety of farm practices, including water management

• Private and public utilities that often run load-management programs to encourage farmers to reduce water demand during peak seasons and hours.

Specific steps that these organizations might take include the following:

• Provide farmers with information about gypsum blocks and the soil-moisture method for improving water distribution and irrigation scheduling

• Conduct field demonstrations of the soil-moisture method in many different locations in the West so farmers can observe and adapt the method on a variety of crops, soils and climatic conditions

- Conduct classes (for example, during winter months) for surface irrigators, who are often hired labor. These classes could show how familiar irrigation practices can be used more effectively when guided by soil-moisture data.

Farmer Nancy Lowe Barker installs a gypsum-block station at INFORM's 1985 educational field day. The 1984 to 1986 field demonstrations showed the benefits of the soil-moisture method and encouraged farmers to use it more broadly.

Part I: Introduction and Findings

These educational steps would be most effective if backed by independent research to help farmers compare the commercial benefits of the soil-moisture method with those of other available water-management methods, and to help them compare gypsum blocks with other soil-moisture monitoring tools. The other most common water-management methods include advance-recession procedures used by technicians to evaluate irrigation efficiency, and water-budget methods used to predict crop water consumption and plan irrigation schedules. (See Chapter 4.) The other soil-moisture monitoring tools used by farmers are primarily the soil auger, tensiometer and neutron probe. (See Chapter 5.)

Finally, there is a need for technical standards and independent technical assessments of the uniformity and other features of gypsum blocks marketed by several small U.S. producers, including INFORM's technical consultant. Each make of block performs somewhat differently. To date, farmers have had no way to learn what qualities they should seek in "good" blocks or which commercial products meet these criteria.

Part II
Background

3
Western Irrigation and Its Problems

From 1945 to 1974 irrigated acreage in the 17 western states doubled, expanding at a rate of 1 million acres a year from 1945 to 1950, and at half that rate from 1950 to 1974. Currently 45 to 50 million farm acres are irrigated. This irrigated land constitutes about a fourth of total farm acreage in the West, but produces about one half, by value, of western crop and animal products.

About 85% of the West's freshwater supplies are used for farm irrigation. Included in this amount is not only the water consumed by crops, but water that evaporates or sinks into the ground while being conveyed to, applied to, or drained off, fields.

Estimates by federal agencies of annual water use on western irrigated fields vary from 83 million acre-feet (maf) to 120 maf, not counting losses during storage, transport or drainage. (One acre-foot equals 325,851 gallons and is roughly the amount used annually by a family of five.)

Despite significant disagreements about volume, however, federal agencies concur that about 60% of the total irrigation supply comes from surface sources such

Part II. Background

About 60% of the West's irrigation supplies come from surface sources such as rivers, lakes and the 444-mile-long California Aqueduct, the West's biggest conveyance, that carries state and federal water supplies to farmers on the west side of California's San Joaquin Valley.

as rivers, lakes and reservoirs. Of this amount, one-third is supplied by projects built by the U.S. Bureau of Reclamation, one-quarter by on-farm sources such as rivers, and the rest by state and local water projects and agencies.

The remaining 40% of western irrigation water is pumped from wells. Between 1945 and 1975, farm pumping rose from 11 to 56 maf per year and supplied most of the acreage brought under irrigation during this period. About 37% of groundwater withdrawals occur in the two states with the most irrigated acres, California and Texas. Another four western states, Nebraska, Idaho, Kansas and Arizona account for 25% of the total.

What Happens to Irrigation Supplies

Water used in western irrigation is never actually used up, although some of it may become unavailable to future western farming through evaporation and transpiration, through contamination, or because it runs off into the ocean. Water is moved around in complex fashion through air, soils, and surface streams and aquifers. The drawing on pages 22-23 depicts the journeys and destinations of irrigation supplies.

Problems of Water Scarcity and Water Quality

The large volumes of water withdrawn for western irrigation from surface streams and aquifers create various kinds of seasonal water scarcity and long-term depletion. In addition, farm drainage returning to surface or underground water sources, and carrying with it salts, naturally occurring trace metals and/or farm chemicals, degrades water quality in various ways and degrees.

Problems of water quality are linked to those of water depletion. Drainage that pollutes a pure water source reduces the amount of usable supply. Stream flows already diminished by dams, irrigation withdrawals or summer drought, may insufficiently dilute farm runoff, and river channels may become little more than drain ditches. The reduced volume of an overpumped aquifer increases the deteriorating effect of salts and minerals leached into it from farm fields.

Turbine pumps lift 40% of the West's irrigation supplies from aquifers to field surfaces. A 120-foot well casing with its center shaft has been withdrawn and dismantled in segments for repairs. When the casing and shaft are replaced in the ground and the pump is running again, water will surge to the surface through the large outer casing.

Groundwater Depletion. Large-scale pumping that depletes groundwater causes the most severe kind of western water scarcity and the one affecting the greatest number of farmers. Each year the volume of water pumped from farm wells but not replenished by rainfall or other means is one-third larger than the annual flow of the Colorado River. The U.S. Department of Agriculture estimates that overpumping for irrigation is causing water tables to fall beneath 16 million western farm acres.

Falling water tables and a tripling of energy prices since the mid-1970's have already reduced pump-irrigated acreages in some areas. Irrigation in the Texas High Plains was abandoned on 1 million out of 6 million acres between 1974 and 1983.

In California's agricultural San Joaquin Valley, where overdrafting averages between 1.5 and 1.8 maf per year, land subsidence of up to 20 feet occurred from 1917 to 1967 over an area of 2,500 square miles. Costs of repairing the damage caused by subsidence to 25 miles of federal irrigation structures amounted to $3.7 million in the 1970's.

Salts. Salts are by far the most common elements carried by farm drainage into water sources, and salinity problems are endemic to irrigated agriculture. The

Part II. Background

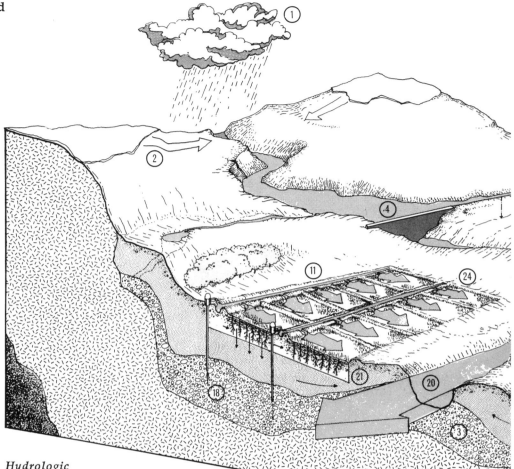

Irrigation and the Hydrologic Cycle

Water for irrigation originates from rain, snowmelt or aquifers (1,2,3). Rain and snowmelt are stored behind dams (4) or diverted directly from rivers (5). Aqueducts and canals (6) are used to carry water from reservoirs and river channels to farm ditches or pipelines (7). During its storage in reservoirs and during its transport through aqueducts, canals and ditches, part of the irrigation supply evaporates (8). Also, some water leaks into the ground through unlined conduits and spills out of conveyor systems (9).

Over 75% of the water reaching western fields is applied directly to gently sloping surfaces. The most common field layouts for surface irrigation are furrow systems (10) and border-strip systems (11). About 20% of water reaching fields is applied by sprinklers (12). A small percentage of water used on fields is applied by drip or trickle systems (not shown).

Much of the water applied to farm fields is used by crops. It is either stored in plant leaves, stems and fruit (13) or transpired into the atmosphere through the "breathing" process essential to plant life (14). Some of the water evaporates from field surfaces (15), some flows off field ends (16) or percolates into deep soil

22

Western Irrigation and Its Problems

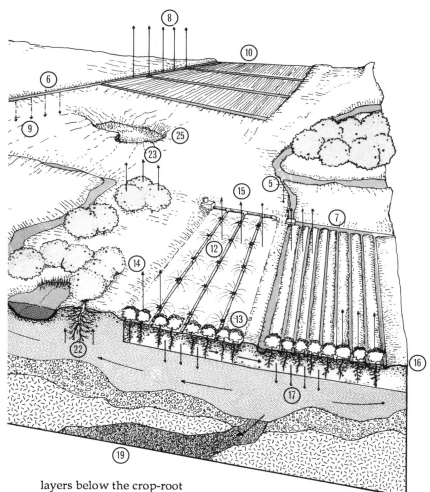

surface stream that flows to the ocean (20). As it travels, some of this water passes through crop-root zones on fields other than where it was applied (21), or through the root zones of non-agricultural plants (such as cottonwood trees) (22). In either case some of this water is then stored in plants and some is transpired (23).

The portion of the irrigation supply that runs off farm fields and remains on the surface eventually reaches one of three destinations. It flows or is repumped onto other farm fields (24); it flows into a stream or river and eventually to the ocean (20); or it empties into a land-locked salt sink (25). Throughout its journey some portions evaporate and others sink into the ground and supply agricultural and non-agricultural plants and animal life.

layers below the crop-root zone (17).

All of the evaporated or transpired moisture eventually returns to the earth as rain, mist or snow, but this may occur a long way from the irrigated fields.

The portion of the irrigation supply that sinks into the ground eventually reaches one of three places: an aquifer from which it can be repumped (18); an underground site from which it cannot be recovered (19); or a major constituents causing salinity are calcium, magnesium, sodium, bicarbonate, chloride and sulfate.

Farm runoff adds 3 million tons of salts a year to the Colorado River. This is about one-third of the system's total salt load (half comes from natural sources). On reaching Mexico, the Colorado's salt levels are about double the Environmental Protection Agency's (EPA) drinking-water standard. Every other western river but the Columbia also suffers from salt problems to some degree.

A second kind of salt problem occurs when salty farm drainage is trapped under fields by impermeable clay shelves close to the surface. About 400,000 acres in California's San Joaquin Valley suffer from this condition. As the undrained water accumulates, it rises into the root zone and damages crops. State sources predict that more than a million acres may be similarly afflicted by the year 2080.

Selenium. Selenium and other toxic elements such as boron, cadmium and arsenic that occur naturally in desert soils, also pollute irrigation drainage and degrade water sources. In California, selenium-tainted drainage from 42,000 farm acres devastated the Kesterson National Wildlife Refuge northwest of Fresno. In a few years it caused the most severe bird deformities on record, as well as killing plants and fish and poisoning cattle. In 1985, the U.S. Department of the Interior closed the 1,200-acre refuge. Then the Westlands Water District, the federal contracting agency that supplies irrigation water to nearly 600,000 farm acres on the west side of the San Joaquin Valley, including the 42,000 causing the problem, plugged the drains leading into the 82-mile-long segment of an unfinished master drain emptying into the Kesterson.

Selenium concentrations capable of triggering fish and wildlife damage also occur in farm drainage near federal refuges in Arizona, Idaho, Montana, New Mexico, South Dakota and Utah.

Pesticides. U.S. farmers, including irrigators, use tens of thousands of pesticide formulations containing about 200 active ingredients. California farmers annually account for about one-tenth of the total U.S. farm pesticide applications.

Only fragmented and sporadic evidence exists to show the kinds, locations and amounts of farm chemicals washed into surface and groundwater. Specific testing and definition of a single problem of groundwater contamination can cost $25,000 to $50,000 according to the EPA. These high costs, as well as scarce time and talent and unproven technologies, discourage preventive and corrective activities, according to the EPA.

Salt damage, appearing as white streaks in an alfalfa field, is common on fields in the southern and western San Joaquin Valley where irrigation water adds up to 4 tons per acre per year. The salts remain when the water is used by crops or evaporates.

Lessons From a Drought

Western farmers would gain financial benefits while also reducing regional water problems, if they irrigated their crops only when and in the amounts needed for good yields. Groundwater overpumping would decline. Farm drainage would shrink in volume. The leaching of farm chemicals into streams and wells would be reduced.

No one knows how much western farmers could cut back their water use without depressing yields, but data on record for the richest farm region in the U.S. suggests the extent. In California's San Joaquin Valley in 1977, during the worst drought in the state's history, farmers on over 5 million irrigated acres used 18% less water than normal. Yet yields were above average for some crops—cotton, for example—and no dollar losses were

Part II. Background

specifically attributed by state sources to the effect of fewer or lighter irrigations.

Farmers who responded of necessity to California's drought could only guess how their forced irrigation cutbacks were linked to crop production. They lacked a method for identifying and repeating, during later seasons, patterns of moisture supply to root zones that had produced equivalent or higher yields during the drought compared with earlier, wetter years. INFORM's technique for seeing into the soil with gypsum blocks provides farmers with such a method.

4
What the Field Surface Doesn't Show

*F*armers who flood all or parts of field surfaces manage three-quarters of the West's irrigated acreage. If they matched their irrigation practices to each field's unique soil types, they could refill crop-root zones more evenly and take an essential step toward greater efficiency of water use. But they cannot see beneath their soils to learn which of their practices may be causing problems of uneven water distribution or which changes might reduce these problems.

The Root-Zone Reservoir

An irrigated field is, among other things, a water reservoir for a growing crop. If the field comprises 80 acres, its reservoir normally measures about half a mile long, a quarter of a mile wide and from several inches to many feet deep, depending on the crop's rooting depth.

Soil textures and soil conditions determine how much water can be stored in a root-zone reservoir. They also determine the rate at which water sinks into the soil (infiltration) and flows through it (percolation).

Part II. Background

Swamping an alfalfa field will not force more water into its soil or store more of it there than irrigating slowly over a longer time. Therefore, farmers seeking to cut back waste must adjust the length, rate and duration of their irrigations to match the unique absorption rates and storage volumes of each field's soils.

Sandy soils normally absorb water easily and drain it quickly. They often store less water than clayey soils, but make a larger fraction of it available to plants. As a rule, they must be refilled frequently to maintain crop growth.

Clayey soils absorb and drain water slowly. They usually store more water than sandy soils, but make a smaller fraction of it available to plants. They typically retain their water supply longer than sandy soils and need refilling less often.

However, fields are not simply sandy or clayey. The U.S. Department of Agriculture classifies soils in 12 major categories. Nature contains countless variations in each category. A single field often has two or more soil types. These may be laid down in spots, streaks, swirls or layers. Salts and other elements may be distributed throughout in ways that modify the soils' interactions with water. No two fields' soils are the same, hence each field's plumbing and reservoir system is unique.

Irrigators can decipher some of a field's soil characteristics by watching what happens to water on the surface. It may sink in quickly on some sections and pond on others. It may dry out quickly in some places, but never dry out in others.

Yet what happens to water in the crop-root zone remains a mystery. Some of it may be trapped by hard layers close to the surface. Some may slip into sandy streaks and speed far below the root zone. Or, in subtler ways, water percolating through the root zone may resist passing between two soil types until a sufficient mass builds up in one of them to break across the boundary line.

Even a soil of uniform texture absorbs and stores water by layers, from the top down. Thus it is possible for an irrigation to saturate a field's top 12 inches, but never reach its deeper regions.

Uniform Water Distribution

The invisibility of the root zone prevents irrigators from seeing whether they refill the crop water reservoir fully and evenly with each irrigation. Doing this job well is a precondition to scheduling irrigations efficiently.

An uneven irrigation wets some sections of a crop-root zone, but leaves other sections relatively dry. (See drawing on page 31.) Even though the wetter sections may be swamped, and water may flow through them to be "wasted" beneath the root zone, cutbacks in irrigation frequency or volume will likely cause or increase parching and risk yield loss on the drier sections. Thus practices causing uneven irrigation must first be corrected before too frequent or too heavy irrigations can be safely reduced.

A Surface Irrigator's Choices

Surface irrigators make many choices about diverse practices that affect the rate, length, duration and timing of water flows across field surfaces. Such choices also affect water-distribution patterns in the root zone beneath the surface, and include:

• The rate of water application through siphons, hydrants (alfalfa valves), or gated pipelines

• The shape, width and spacing of furrows used to convey water to row crops, and the width of the broad basins or checks used for flooding alfalfa, close-grown grains and many orchards

A pit dug into the ground reveals that irrigation water, trapped by an impermeable soil layer below the pit, waterlogs the lower soil band (darker color), and confines crop roots to the drier (lighter) band close to the surface. If farmers could seen beneath their fields without digging holes, they would often discover odd wet and dry patterns that could help them assess and improve their irrigation practices.

Part II. Background

Siphons are being withdrawn from one part of a supply ditch along a sugar-beet field to be reset in another part of the ditch. Farmers often reset their siphons five or six times and leave them in place for 10 to 12 hours in order to irrigate a field over several days.

- The kinds of tillage, leveling and treatment methods used to plow land, break up hard soil layers, mix different soil types, fill in low spots and reduce clogging caused by salts' interactions with soils

- The slope and length of the run from the side of the field where the water is applied to the side where it drains off.

If farmers could detect problems of uneven water distribution in crop-root zones they could discover how to reduce these problems by changing their practices. They could first try changing the rates of water application and later try more time-consuming or complex remedies for problems meriting the additional effort and expense.

What the Field Surface Doesn't Show

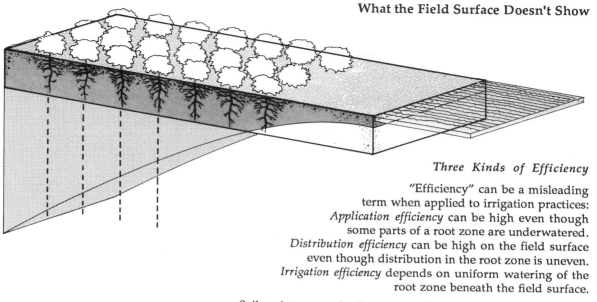

- Crop root zone
- Runoff
- Overirrigated portion
- Underirrigated portion
- Percolation beneath root zone
- Water leaching salts out of root zone

Three Kinds of Efficiency

"Efficiency" can be a misleading term when applied to irrigation practices: *Application efficiency* can be high even though some parts of a root zone are underwatered. *Distribution efficiency* can be high on the field surface even though distribution in the root zone is uneven. *Irrigation efficiency* depends on uniform watering of the root zone beneath the field surface.

Soil-moisture monitoring at several field locations and depths is a practical way for farmers to spot-check distribution patterns in the root zone and use this information to work toward greater irrigation efficiency.

Water losses. Runoff plus deep percolation.

Overwatering. The amount of water applied in excess of that needed to refill the entire root zone.

Application Efficiency. The amount of water stored in the root zone as a percentage of the total amount of water applied to the field.

Distribution Efficiency (Uniformity). The uniformity of depth of water applied to the field surface. Such uniformity does not, however, indicate patterns of distribution beneath the surface.

Irrigation Efficiency. The water stored in the root zone plus the water used to leach out salts as a percentage of the total amount of water applied to the field.

Two Irrigation-Management Methods That Serve Other Needs

Advance-recession and water-budget methods are two common means used by technicians and consultants to provide farmers with useful information for improving

Part II. Background

An open hydrant spills water from an underground pipeline onto an alfalfa field. There is no practical field tool available for measuring water outflow through hydrants.

their irrigation efficiencies. Yet neither method gives irrigators a field-level tool for doing their own investigations and corrections of distribution problems.

Advance-recession methods are one-time, day-long evaluations of the efficiency of surface-irrigation practices that give farmers a starting point for making changes on a particular field. The procedures involve careful measurements of the time intervals at which an irrigation stream advancing down a furrow or check reaches a series of checkpoints. These time-distance data are then plotted on a graph to produce an "advance curve." The drying-out of a furrow or check after the water is turned off can similarly be clocked from point to point and graphed. Data on water applications, soil-intake rates and runoff volumes can also be analyzed, as can data on soil-moisture levels. However, these meth-

What the Field Surface Doesn't Show

Gates on the far side of this pipeline discharge water into furrows on a field that will soon be planted with cotton. Once this pre-irrigation is over, the pipeline will be dismantled and trucked off for use on another field.

ods do not systematically incorporate subsurface data with measurements taken on the field surface.

Weather-data methods are used to help farmers estimate crop water consumption and plan irrigation schedules. They are most effective on fields where distribution problems have been previously analyzed and reduced.

From continuously gathered measurements of solar radiation and other weather variables, the water used by plants (usually grass) surrounding weather instruments is calculated. These "evapotranspiration" (ET) data are mathematically adjusted for farm crops located miles from the weather station. (The use of fieldside weather units obviates the need for this complex adjustment of values.)

On a field managed with either crop-adjusted or locally gathered ET data, the crop-root zone, whose

Part II. Background

Well-designed and managed sprinklers irrigate field surfaces very evenly. However, only soil-moisture monitoring shows farmers whether water is stored evenly beneath the surface.

water-storage capacity is determined, must be fully refilled at planting time. Then ET values are added daily until they total an amount judged to have depleted root-zone moisture to levels low enough to require an irrigation. A watering judged sufficient to refill the reservoir is applied, and the process begins again. However, these methods do not specify the locations or frequency of soil-moisture monitoring needed to determine that a root zone is full at the beginning of the season and to assess the effectiveness of formula-based scheduling decisions later on.

Part III
Tools and Methods

5
Gypsum Blocks

*I*NFORM's 1984 to 1986 field tests with gypsum blocks showed how systematically gathered soil-moisture data enable surface irrigators to see the results of their practices on crop water-supply levels in the root zone and adjust their practices to achieve more efficient irrigation. Gypsum blocks are the most practical soil-moisture tool available for tracking water-distribution patterns at multiple field sites and depths as needed to assess and improve irrigation uniformity. INFORM chose gypsum blocks for its tests because of their advantages over three other tools for assessing soil-moisture levels: the soil auger, tensiometer and neutron probe.

The gypsum block INFORM used is a marshmallow-size plug made with plaster of paris. Two electrodes embedded in it are connected to two insulated wires. As the block is buried, the wires are drawn to the surface for testing with a battery-powered impedance meter (AC resistance).

The impedance meter indicates the amount of electrical current flowing through the circuit formed when the buried block's electrodes are connected through

Part III: Tools and Methods

Graph 1. How to Interpret Gypsum-Block Readings

The graph plots hypothetical meter readings taken on dates 1 through 14 from three gypsum blocks buried at 1-foot intervals beneath the surface.

1) A reading of 94 to 96 shows *soil saturation*. More water is temporarily draining through the soil than can be stored in its pores. All three depths are saturated on dates 1 and 2.

2) A reading of about 90 shows *field capacity*. The soil is storing as much water as possible after the excess has drained off. This occurs on dates 3, 5 and 8 for the respective depths.

3) Steeply falling curves show *rapid water uptake* by plant roots. This occurs between dates 3 and 6 at the 1-foot depth and on later dates at the other two depths.

4) Flattening curves at the low end of the scale show that the *soil moisture is nearly depleted*. This occurs after date 6 at the 1-foot depth and after date 11 at the 2-foot depth.

5) A reading of 3 shows the *permanent wilting point*. At this point of extreme soil dryness, the roots of most agricultural plants can extract no more water. This point is reached on date 14 at the 1-foot depth.

Key		
	1 ft	•
	2 ft	+
	3 ft	✶

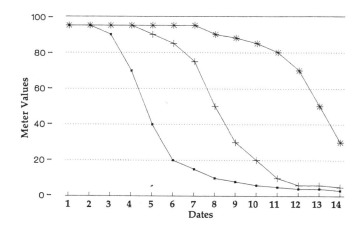

the wires to the meter. A buried block absorbs and releases moisture similarly to the surrounding soil. As moisture levels fall in the surrounding soil and block, the forces exerted by the soil and block particles to hold onto the remaining water increase. As these forces increase, electrical conductivity through the block decreases.

What the Meter Readings Mean

Changes in impedance to electrical current flowing through buried gypsum blocks show *relative* changes in soil-moisture levels. *They do not indicate the volume of water stored in the soil. Nor do they reveal percentages of soil-water depletion.* Even so, block data are an effective guide to improving irrigation practices and increasing water-use efficiency on all common soil types.

The meter used with INFORM's blocks registers soil-moisture changes on a scale of 1 to 100. Higher readings indicate higher moisture levels. The simple rules for interpreting the scale's numbers are illustrated by reference to the graph of hypothetical readings above. These rules are the same for all soil types.

Technical Features of Gypsum Blocks

INFORM used one of several designs of gypsum blocks produced commercially. These designs are comparable in many ways, with the key exception of *uniformity*.

Uniformity. Each gypsum block used by INFORM registers the same meter value between 1 and 100 (for all practical purposes) at the same moisture level over the lifetime of the block. All blocks are highly uniform in this respect both within and among production lots. Uniformity is achieved by using high-quality gypsum, by good electrode design, by carefully controlling manufacturing and by curing the finished blocks through wetting and drying procedures.

The uniformity of the gypsum blocks used by INFORM has been documented in laboratory tests conducted by D.W. Henderson, a retired professor from the University of California at Davis. Dr. Henderson's research has also revealed considerable differences in uniformity among commercially produced blocks. (See Appendix F for an evaluation by the Westlands Water District field staff of the uniformity in field applications of the blocks used by INFORM.) However, there is no technical standard for assessing the uniformity of commercially produced blocks, nor is there any accepted test procedure for evaluating this critical aspect of their performance.

Sensitivity. Most gypsum blocks currently available are sensitive over the soil-moisture range from "field capacity" to the "permanent wilting point" in all major soil types.

Durability. All gypsum blocks gradually dissolve in water. Thus in continuously wet soils they may last only

A half-constructed gypsum block (right) shows the two concentric electrodes to which two insulated wires will be attached before a second pouring of gypsum and water completes its construction. The finished block (left), which is 1-1/4 inches in diameter, is buried in a crop-root zone with its two insulated wires drawn to the surface for testing with an impedance meter.

one season. However, in well-drained irrigated soils they can last up to five years.

On most surface-irrigated fields planted with row crops, one season's durability is all that is required. Gypsum-block installations are destroyed when these fields are plowed under at the end of each season unless special precautions are taken.

By contrast, on surface-irrigated alfalfa fields producing hay for three to five seasons after a single planting, and on fields of other perennial and permanent crops (like orchards and vineyards), block installations often last several seasons in well-drained soils.

Some gypsum blocks are impregnated with non-rotting fibers, usually glass or nylon. These units last longer than blocks made with gypsum alone.

The Buffering Effect. The impact of dissolved salts on readings from blocks consisting of gypsum and water alone is negligible throughout the range of saline concentrations commonly found in agricultural soils. Calcium ions in solution from the gypsum (calcium sulfate) neutralize the impact of salts on the electrical conductivity of a soil solution. Even very high salt concentrations affect gypsum blocks only when the soil is near saturation. However, readings from nylon- or glass-impregnated blocks are more affected by salts in the soil solution.

Responsiveness. Gypsum blocks respond quickly to changes in moisture levels in the surrounding soil. When first installed, however, it usually takes two days for the blocks' moisture levels to reach equilibrium with the soil's moisture levels. By contrast, lag times of several hours or a day or two between an irrigation and block reactions are due primarily to the block's depth and the different rates at which water percolates through different soils, rather than to the responsiveness of the block itself.

Practical Features

The chief practical attractions of gypsum blocks are their low cost, permanent siting, ease of use and adaptability.

Cost. INFORM's gypsum blocks cost $5 apiece. Other blocks cost from $3.50 to $9.50 apiece. This low cost per unit allows many units to be used economically on the same field, which is an essential requirement for tracking water-distribution patterns.

Permanent Siting. Gypsum blocks remain in place for at least one growing season, and readings from them can be gathered at any time. This permits continuous monitoring of the same field sites, a second requirement for tracking water-distribution patterns.

Ease of Use. Farmers can easily install and take readings from gypsum blocks. Installation procedures used by INFORM are described in Chapter 6.

Adaptability. Gypsum blocks can be used on all soil types and all irrigation systems, but with this qualification: When root zones are maintained at high moisture levels, gypsum blocks tend to dissolve quickly. Thus, they are usually impractical for use a) on drip-irrigation systems where a limited volume of soil is kept continuously wet, and b) on fields that are frequently and heavily irrigated.

In addition, on very sandy soils, such as those found in southern California and Arizona, gypsum blocks must be read more often than on silty or clayey soils, to keep track of the more quickly falling moisture levels between irrigations in these soils.

An impedance meter attached to a block's wires sends an electric current through the block. High readout numbers on an arbitrary scale of 1 to 100 indicate that more current is passing through a block due to wet conditions in both the block and the surrounding soil than occurs when the block and soil are dry.

Part III: Tools and Methods

The use of gypsum blocks to manage sprinkler-irrigation systems is not described in this report. However, many of the same basic principles of monitoring and testing used on surface-irrigated fields can also be applied to sprinkler systems.

Three Other Tools for Monitoring Soil Moisture

Three other tools used for measuring soil-moisture levels on irrigated farm fields are the soil auger, the tensiometer and the neutron probe. For both technical and practical reasons, none of them offers farmers as many advantages as gypsum blocks.

The *soil auger* removes a soil core for hand testing by squeezing, rolling or sifting. The different "feel" of major categories of soil textures, at different degrees of wetness, can only be learned through experience.

The subjectivity of such interpretations renders soil augers technically inferior to gypsum blocks, tensiometers and neutron probes. In addition, a farmer using an auger never probes a field twice in precisely the same place, so reliable continuous monitoring of even one site is impossible. Finally, the auger cannot penetrate dry soil layers that harden at the surface between irrigations in order to sample lower soil depths that may still be wet enough to sustain crop growth for days or even weeks.

The *tensiometer* is a plastic tube of water with a vacuum gauge at one end and a ceramic tip at the other. The ceramic tip and part of the tube are buried and the vacuum gauge remains on the surface. When wet soil begins to dry out, root-suction pressure increases, pulling the tube's water through the ceramic tip, and creating a vacuum in the tube. The strength of the vacuum is equivalent to the root-suction, hence the vacuum gauge indicates soil-moisture levels.

Each tensiometer, like each gypsum block, measures moisture levels only at one location and one soil depth. Therefore, several tensiometers must be used to probe several depths at one location simultaneously to track water-penetration and depletion patterns.

The $30 to $40 cost of a tensiometer makes multiple-site monitoring with them very expensive. Because tensiometers' vacuums break at relatively high moisture levels in high-storage soils, their use can encourage unneeded irrigations. Labor costs for installing tensiometers are lower than for gypsum blocks, but overall labor costs are higher because tubes must be refilled with distilled water each time their vacuums break. This often occurs several times during a growing season. Yet the decisive limitation of these tools for growers of field crops is that they stick up above the ground and are destroyed by (and can do damage to) farm machines.

The *neutron probe* is a tool the size of a small suitcase, with a flexible probe unit attached to it. Through the probe, neutrons from radioactive material contained in the body of the instrument are emitted into soil that has been previously analyzed for its water-holding capacity and prepared for testing. ("Access tubes" are installed at probe sites.) As fast neutrons from the probe collide with hydrogen atoms in soil water, they slow down. A counter records the number of neutrons per volume of soil that have been retarded in this fashion. This count establishes the number of water molecules in the tested soil area, and thus its water content.

An area several feet deep and about 10 inches in diameter can be probed at one access site. The probe unit itself is carried from site to site for the testing. Because neutron probes contain radioactive material, their

Several models of gypsum blocks are produced commercially by small-scale manufacturers, while some blocks, like the long, banded type, are used primarily for research. Blocks vary in uniformity over time and among production lots, yet no uniformity standards have been developed for evaluating this important aspect of their performance.

operators must obtain licenses from the Atomic Energy Commission.

Neutron probes are rarely used by farmers because of their technical complexity, licensing requirements and high cost.

The chief characteristics of all four soil-moisture-monitoring tools are summarized in Table 1. Because of their various limitations, augers, tensiometers and neutron probes are generally used at only one or a few field locations to test soil dryness before irrigating. However, unless the fields have been analyzed for water-distribution patterns, it is largely a matter of luck if the sites chosen for monitoring are optimal. Only the gypsum block is a practical tool for the continuous, multiple-site monitoring needed to identify and improve water-distribution patterns. Only when these patterns are known and improved where possible, can an efficient irrigation schedule be chosen.

Why Gypsum Blocks Have Been Overlooked

Despite their practical and technical advantages over other tools used to measure soil-moisture conditions on surface-irrigated fields, gypsum blocks have not to date been widely adopted by farmers. This may result from three possible causes.

First, little public or farm-level attention was devoted to the efficient use of farm-water supplies until the 1970's. Before then, it was widely assumed that the West's future water needs could be met by continually expanding supply through the construction of new dams, reservoirs and aqueducts.

Second, surface irrigation is often seen as inherently inefficient compared to high-tech sprinkler and drip

TABLE 1. FOUR TOOLS FOR MONITORING SOIL MOISTURE

Technical considerations

Gypsum blocks @ $3.50 - $9.50

1. All modern gypsum blocks are sensitive throughout the soil-moisture range from field capacity to the permanent wilting point.

2. Blocks made of gypsum buffer the effects of salts in soil water except in highly saline conditions which slightly distort readings in the wet range. Glass or nylon fiber units provide less buffering.

3. The uniformity of the gypsum blocks used by INFORM is high, but uniformity varies among other blocks.

Soil auger @ $50 - $100

1. Subjective judgments based on the feel of damp soil form the basis of soil-moisture assessments.

2. With experience, farmers can learn to judge soil-moisture by feel and apply this knowledge usefully to their irrigation-scheduling decisions.

Tensiometer @ $30 - $40

1. Tensiometers are reliable in the wet range of all soils. However, their vacuums break when their strength equals the pressure of the earth's atmosphere. This occurs in the wet range for many common soil types. Thus this tool tends to encourage over-irrigation.

Neutron Probe @ $3,500

1. The neutron probe is the best device available for measuring soil-water volumes precisely and is thus an excellent research tool. However, in field situations where costs are of concern, the complex and time-consuming procedures for recalibrating the instrument for each field's different soil type(s) are often ignored. Instead, readings are interpreted as showing higher or lower soil-moisture levels but not volumes of water stored in the soil.

Practical considerations

1. In general, blocks are effective on all soil types and all but drip irrigation systems.

2. Gypsum blocks are easy to install and read, and remain "permanently" in place throughout the growing season. They require no maintenance.

3. Gypsum blocks are not effective on continuously wet soils because they dissolve.

4. Blocks need to be read very frequently on very sandy soils.

1. Augers never probe exactly the same site twice in a row and thus cannot produce a reliable, continuous record of moisture changes and distribution patterns at particular field locations and depths.

2. An auger cannot be forced through a hard, dried-out topsoil layer to reach a moist zone which may lie beneath.

1. Tensiometers are easier to install than gypsum blocks.

2. Tensiometers are most frequently used in orchards. They are impractical to use on field crops because they stick out of the ground and are destroyed by (or cause damage to) farm machines.

3. Every time its vacuum breaks, a tensiometer tube must be refilled with water. This is a time-consuming maintenance process.

1. The time required to use neutron probes effectively limits their use to one or two sites per field.

2. Operators must be licensed by the Atomic Energy Commission.

irrigation, despite good research showing otherwise. Moreover, the search by engineers and others for sophisticated (and usually expensive) new irrigation equipment has attracted vastly greater efforts than the more modest and still undone job of finding efficient management methods for farmers to use in analyzing and reducing distribution problems on surface-irrigated fields.

Finally, very little information about gypsum blocks has been made available to farmers or agriculture students since the blocks were first introduced in the 1940's. INFORM could find only one article describing step-by-step procedures for using gypsum blocks to manage irrigation scheduling. In 1951, *Agricultural Engineering* described large-scale experiments with gypsum blocks on a Hawaiian sugar plantation where "two men managed nine separate fields, widely scattered over the plantation, in approximately one-fourth of their working time." This was written before the water-cost reductions and crop yields associated with the gypsum-block method were analyzed. Yet, the article concluded, "the savings to date in terms of irrigation labor cost has more than paid for installation and operation costs."

According to Wes Robbins of the U.S. Soil Conservation Service in Colorado, some significant field work with gypsum blocks in a number of western states over the past decades has never been published or shared.

For whatever reasons, the potential value of gypsum blocks has never been adequately demonstrated to the tens of thousands of western farmers who could benefit by using them.

6
Installing Gypsum Blocks on Farm Fields

*I*NFORM installed 12 sets or "stations" of gypsum blocks on fields with typical runs of 1,200 to 1,250 feet from the water source to the drain ditch. (Four fields had longer runs and two had shorter runs.) The four blocks in each station were buried at depths of 1, 2, 3 and 4 feet. Installations were usually made in damp soil early in the growing season.

From 30 to 45 minutes were required to install one four-block station. One full day was usually sufficient for installing all 12 stations on a typical field.

Gypsum Blocks Must Be Surrounded by Rootlets

In order to register changes in soil-moisture levels caused by the plant growth, gypsum blocks must be located in the plant-root zone where they are surrounded by rootlets. Thus, on fields of annual row crops such as cotton and tomatoes, that grow from seed to maturity in a single season, blocks were installed only after young plants had reached a height of 6 to 8 inches.

Part III: Tools and Methods

The Two-Strip Test Design

By monitoring two separate field strips, INFORM was able to compare results of irrigation improvements on its test strip with results achieved on the second (owner) strip managed by the farmer.

TOP, MID and END monitoring sites on each strip tracked variations in water-distribution patterns in the crop-root zone at intervals down the run. A pair of stations at each site yielded two readings at each depth as insurance against the loss of data in the event of problems with any one station or block.

Blocks at the 1, 2, 3 and 4-foot depths—that part of the root zone from which field crops normally derive half or more of their moisture and nutrients—monitored different soil layers whose moisture conditions often vary.

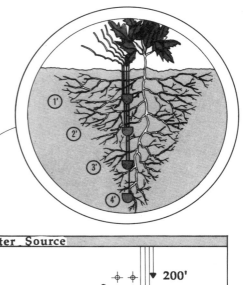

One Gypsum-Block Station

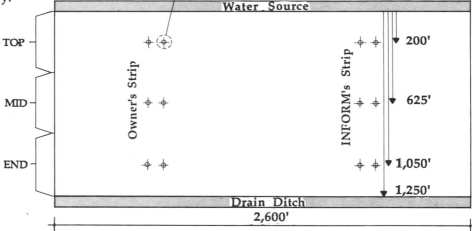

Each station was sited between two adjacent plants in the plant row.

No such precautions were needed when installing gypsum blocks on alfalfa fields. These fields were tested only during their second, third, or fourth years of hay production, after thick-growing beds and deep root systems had been established.

Installing Gypsum Blocks on Farm Fields

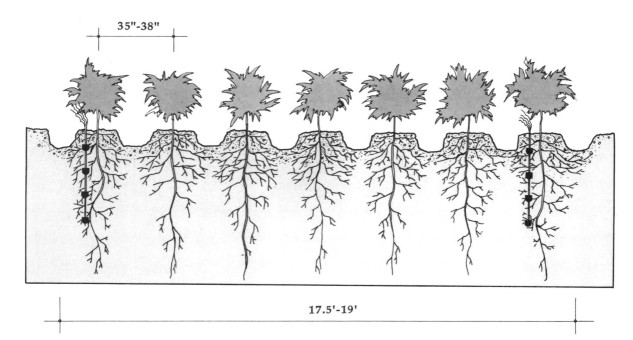

Cross-Section of a Monitored Strip in a Field of Row Crops

One pair of gypsum-block stations spaced 17.5 to 19 feet apart mark the width of a monitored strip in a field of row crops (such as corn, cotton, or sugar beets) where the distance between bed centers is 32 to 38 inches. On 60-inch tomato beds the stations were four beds or 20 feet apart.

The Pattern of Field Installations

INFORM's two-strip pattern for monitoring and testing fields (see drawing, facing page) was developed in the late 1970's by Peter Mueller-Beilschmidt, INFORM's technical consultant. It is a model that farmers can follow precisely or modify to meet special field and crop requirements. Where the soil type is known to be relatively uniform, and/or the irrigation run is short, fewer stations may be used. On crops with shallow root systems, or on deeper-rooting crops where moisture-changes at the 2 or 3-foot depth are used for scheduling irrigations, monitoring to the 3-foot depth can be sufficient. On fields with marked variations in soil types, special distribution problems, long irrigation runs and/or very large total areas, additional monitoring

Part III: Tools and Methods

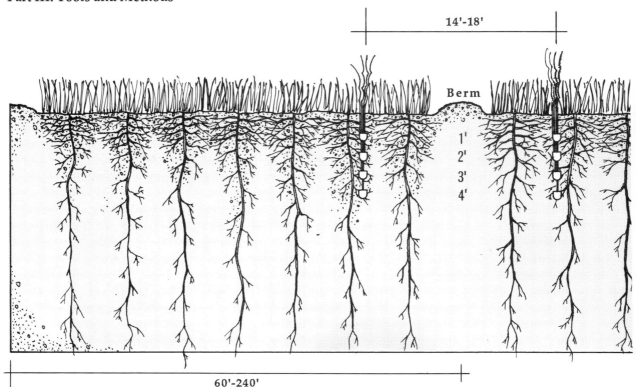

Cross-Section of a Monitored Strip in an Alfalfa Field

One pair of gypsum-block stations buried 14 to 18 feet apart in a monitored field strip straddle the low levy (berm) that separates two adjacent checks or broad beds in which alfalfa is planted and through which irrigation water flows. Due to check widths that varied from 30 to 120 feet, the width of the monitored and tested strips on INFORM's alfalfa fields ranged from 60 to 240 feet.

strips, stations and/or depths may be preferred. However, two basic elements are essential to this model: Two or more sites monitor distribution patterns down the run. Two or more monitored strips permit changes on one strip to be compared with standard practices on the other(s).

Field-strip lengths on INFORM's fields varied with the length of the run from the water source to the drain ditch, and ranged from 500 to 2,500 feet. However, all but six fields had runs of about 1,250 feet.

Field-strip widths on INFORM's fields ranged from 15 to 240 feet, depending on whether a furrow or a border check system was being monitored. (See drawings on

Installing Gypsum Blocks on Farm Fields

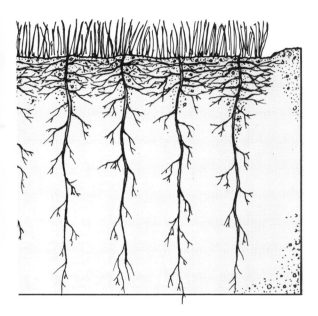

pages 49-50.) However, on any one field the widths of the two monitored strips were always the same.

Installation Procedures

The following photographs illustrate the procedures used for installing four blocks in one gypsum-block station. The site pictured is unplanted because blocks were being buried to monitor a preplanting irrigation (pre-irrigation).

Part III: Tools and Methods

Step 1. *Pre-soaking four gypsum blocks.* Gypsum blocks are soaked in water to assure good contact with the soil, which is essential for effective soil-moisture monitoring. In addition, moisture levels in newly buried, pre-soaked blocks come into equilibrium with moisture levels in the surrounding soil more quickly than with dry blocks, and monitoring can begin sooner.

Step 2. *Digging a 4-foot hole.* Using a 1-1/2-inch-diameter auger and a hammer, a 4-foot-deep hole is dug by extracting 6 to 8 inches of undisturbed soil core at a time. When the soil is too dry for digging down 4 feet, water is poured in the partially dug hole to soften the soil and the installation process can be finished the next day. (See photograph.)

Step 3. *Reserving the soil core.* Soil-core segments are removed from the auger tip and set aside in the order in which they are removed.

Installing Gypsum Blocks on Farm Fields

Step 4. *Checking saturated blocks with a meter.* Each soaked gypsum block is checked with an impedance meter for a "saturation" reading of 94 to 96 to ensure that it is functional before being buried. From one to four knots is tied into the top end of each lead wire to code the depths at which each block is to be buried. (One knot indicates 1 foot of depth.)

Step 5. *Moistening the soil and installing the 4-foot-deep block.* The soil at the base of the 4-foot hole is moistened to ensure good soil-block contact. (See photograph.) Then, while the wire is held in one hand, the gypsum block coded for the 4-foot depth is lowered into the hole. A 6-foot length of 3/4-inch plastic tubing guides it to the bottom and presses it firmly into the damp, loose soil.

Part III: Tools and Methods

Step 6. *Replacing a segment of soil core.* The last-removed segment of soil core is crumbled on top of the block and tamped firmly into place until the hole is filled to the 3-foot depth. A tightly packed soil column prevents water from later irrigations running into the hole and skewing gypsum-block readings. (See photograph.)

Step 7. *Installing three more blocks.* Steps 5 and 6 are repeated for placing the three remaining blocks at the 3, 2 and 1-foot depths in the hole. Segments of the soil core are replaced as nearly as possible in the locations from which they were extracted. Care is taken to moisten soil below and above each block as it is installed to ensure good soil-block contact.

Installing Gypsum Blocks on Farm Fields

Step 8. *Finishing up.* The wire leads from all four buried blocks are gathered and held while the topmost portion of the hole is refilled, using some extra soil. This additional soil is required because soil volume is reduced while tamping down the replaced core.

Part III: Tools and Methods

Step 9. *Flagging the station*. On row crops, (or when testing pre-irrigations, as pictured here) the site of each gypsum-block station is marked with a small colored flag.

Installing Gypsum Blocks on Farm Fields

Step 10. *Burying wires and marking stations.* On alfalfa fields, which are mowed once a month during the growing season, the wire leads from each gypsum-block station at the soil surface are buried in a 6-inch length of 1-inch plastic tubing to protect them from being cut off along with the hay. Also, orange paint is used in place of flags to mark the locations of the stations. The paint is sprayed on the ground around each buried tube containing wire leads, and on the berm nearby. (See photograph.)

However, when stations are used for two or more seasons on the same alfalfa fields, the orange spray paint often washes off over the winter. In order to relocate the stations in the spring, a square-foot patch of alfalfa can be unearthed in the fall at a berm location at paced-off distances between two stations. This spot will remain bare and thus serve as a marker the following spring.

Step 11. *Flagging the test strips.* On the edge of each field monitored by INFORM, colored flags mark the strips along which the gypsum-block stations are buried.

7
Gathering and Graphing Gypsum-Block Data

From 1984 to 1986, INFORM gathered over 40,000 readings from over 1,500 gypsum blocks buried in 32 California fields. These data were used to evaluate farmers' irrigation practices and conduct experiments in improving them. INFORM plotted the data on graphs to show farmers and others what the information revealed and how it was used in tests.

Gathering Data

On a typical 80-acre field, INFORM normally read all 48 blocks used to monitor soil-moisture changes at 12 locations and four depths in 30 to 45 minutes. It took only a couple of minutes to test each station's four buried blocks by connecting their projecting wires to the terminals of the hand-held meter. The numerical readout was recorded on a form, or simply written on a piece of paper attached to the back of the meter to be transferred later to the form.

Gathering and Graphing Gypsum-Block Data

It takes only one or two minutes to "read" a station of four buried gypsum blocks by attaching wires from each block in turn to an impedance meter and recording the readout number. All 12 stations in an 80-acre field can be read in about 30 to 45 minutes.

INFORM normally gathered soil-moisture readings twice weekly from each field. The readings were sometimes spaced farther apart after irrigations when soggy field conditions prevented easy access to monitored sites.

When farmers are first using gypsum blocks to identify moisture-depletion and replenishment cycles on particular fields and crops, they may prefer to take readings more than twice weekly. However, once they master the method, weekly readings are usually sufficient except during flowering and fruiting periods when crops use water very rapidly and delayed irrigations may reduce yields. During such periods readings must often be taken more frequently.

Part III: Tools and Methods

Graph 2. Moisture Changes Between Readings

This graph shows that by July 14, 1986, a July 6 irrigation had completely saturated (meter values of 95) the 1 and 2-foot depths of an alfalfa field-strip, while partially replenishing the 3-foot depth and leaving the 4-foot depth largely unaffected.

The precision with which INFORM's graphs revealed actual moisture changes in crop-root zones depended on the frequency of their collection. On this graph, for example, a dashed extension of the 1-foot curve from July 3 through July 7 to July 14 probably depicts more closely the actual moisture changes occurring just before and immediately after the July 6 irrigation than the line connecting the readings of July 3 and July 14, taken 11 days apart.

Such data gaps most commonly occurred following irrigations, due to soggy field conditions. However, they rarely created practical problems of interpretation. Here, for example, the July 14 readings show that the 1 and 2-foot depths are still saturated a week after the July 6 irrigation and therefore that they did not drop lower during this period.

Key		
	1 ft	-■-
irrigation }	2 ft	+
	3 ft	✶
cutting │	4 ft	▣
To read meter values, see page 38.		

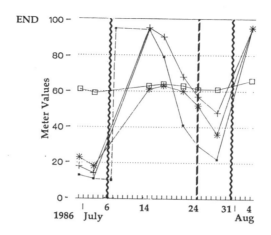

INFORM monitored alfalfa fields for four or five month-long growth cycles during the growing season and collected about 1,300 pieces of data per field. Row crops were monitored throughout their three or four-month growing seasons and about 900 pieces of data were collected per field. However, light waterings used to fertilize and germinate seeds (laybys) were not monitored because block installations during this stage of growth could have damaged fragile young root systems.

Graphing Data

INFORM plotted data gathered from the 48 gypsum blocks monitoring each field on six separate graphs, one each for the TOP, MID, and END sections of INFORM's and the owner's strips. Yet, once farmers learn the soil-moisture method, they can often apply the data directly to their irrigation decisions, without using graphs.

Page 60 illustrates INFORM's graphing technique. The graph plots the *averages* (for each soil depth) of data gathered over a two-month period in 1986 from two four-block stations in an END section of a strip managed by INFORM on an 80-acre alfalfa field.

PART IV
INFORM's Field Work

8
Cooperators and Fields

Using gypsum blocks, INFORM studied 32 fields in California's Central Valley from 1984 to 1986. Apart from one sprinkler-irrigated field, INFORM examined only fields that farmers irrigated by flooding surfaces or portions of them.

In all, INFORM analyzed 2,068 surface-irrigated California acres, a tiny fraction of the state total. Yet INFORM's sample was sufficiently diverse and its results sufficiently encouraging to suggest that much of the surface-irrigated acreage in California and the West could eventually benefit from the soil-moisture method of irrigation management.

Criteria of Field Selection

INFORM made no effort to select typical soil types on the surface-irrigated fields in its study because each field's soil types and conditions are unique. In fact, soils were largely ignored until, after the tests were under way, the soil-moisture method itself could be used to explore them. Farmers, crops, water sources, and field

Part IV: INFORM's Field Work

Tony Barcello of Kings County says that he and his neighbors know that overwatering hurts cotton yields. But he says that some of his neighbors still think they need three irrigations a month to grow good alfalfa.

locations were important in INFORM's choices of test fields.

• Cooperators were selected who were interested in assessing the effectiveness of INFORM's method for themselves. Farmers were recommended as potential cooperators by the U.S. Soil Conservation Service and/or neighboring farmers and were accepted into the study on a first-come basis.

• Only widely-grown and high water-using crops were analyzed so that INFORM's results, if positive, would attract the attention of large numbers of farmers.

• Groundwater-irrigated fields were preferred over fields using surface supplies because the tripling of pump-electricity rates since the mid-1970's has, in general, put farmers using pumped water under the greatest pressure to improve their irrigation efficiency.

• The Central Valley was selected for the study. This 450-mile-long, 100 mile-wide trough, stretching from the Shasta Dam to the Tehachapi Mountains, encompasses 75% of California's irrigated agriculture. Test fields were divided among sites in the Sacramento and San Joaquin Valleys that comprise the Central Valley, in order to demonstrate the method's utility in the Valley's two major geographic and climatic regions. (See map on facing page.)

Cooperators

The 12 farmers who cooperated most closely with INFORM owned or managed 27 of the 32 fields in the study. One of the farmers worked with INFORM for

Cooperators and Fields

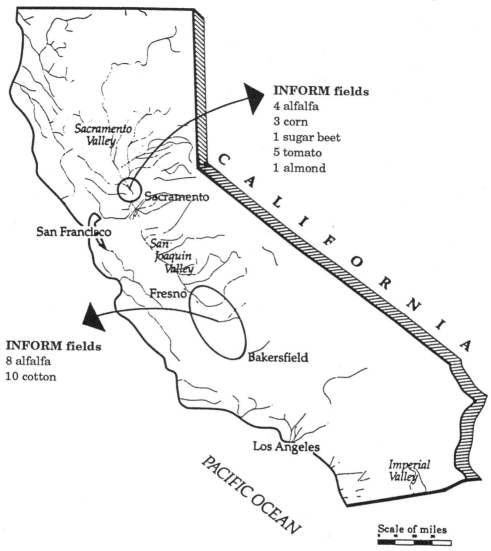

three growing seasons, seven for two seasons and four for one season. (See Appendix A.)

Four of INFORM's 12 principal cooperators were farm managers; eight were owners or part-owners of family-owned farms (some of which were incorporated). Four

Table 2. Crops Studied by INFORM on 32 California Fields

Crop	Number of fields of each crop			Total fields
	1984(pilot)	1985	1986	
Alfalfa	1	9	9*	12
Almonds	1			1
Corn	1	2		3
Cotton		4	7†	10
Sugar beets		1		1
Tomatoes	1	2	2	5
Total	4	18	18**	32

* Includes 7 fields also studied in 1985
† Includes 1 field also studied in 1985
** Includes 8 fields also studied in 1985

of the 12 cooperators farmed in the Sacramento Valley and eight in the San Joaquin Valley. Their farms ranged from 500 to over 2,000 acres.

The youngest INFORM cooperator was 22 and the oldest was 80. They exhibited a wide range of farm experience, management styles and irrigation practices. In general, they were recognized as "good" or "progressive" farmers by their neighbors.

Crops

INFORM analyzed alfalfa, almonds, corn, cotton, sugar beets and tomatoes. All of these crops fall within the ten most broadly planted and highest water-consuming crops in the state.

Nearly 85% of INFORM's tests were conducted on alfalfa, cotton and tomatoes which are planted on about one-third of California's irrigated acreage and use about one-third of the state's annual developed water supply.

Alfalfa is grown everywhere in the state and by nearly every field-crop farmer. Although its earnings ranked it sixth among California's farm products in 1986, its economic importance is even greater than this statistic suggests.

Alfalfa is an essential component in the diet of dairy cows which produce the state's most important farm products, milk and cream. In 1986 the value of California's milk and cream production was $2.06 billion, or about 14.2% of the state's total farm-product earnings.

Also, because alfalfa's deep root structures open up soils and fix nitrogen in them in crop-usable form, it is commonly rotated with other field crops such as corn or cotton to maintain good soil conditions and soil fertility.

Cotton and tomatoes are the two highest earning row (non-specialty) crops in the state. Cotton is planted principally in the southern San Joaquin Valley and in the Imperial Valley. Tomatoes are planted in most of the state's growing regions, but principally in the southern Sacramento Valley and on the west side of the San Joaquin Valley.

Water Sources

On 23 of INFORM's 32 fields, irrigation water was pumped from underground wells. On the remaining nine, water was delivered to the farm by a nearby water district which in turn received its supply from a federal water project. On six of these nine fields, pumped water was also used occasionally.

Part IV: INFORM's Field Work

TABLE 3. MAJOR SOIL TYPES FOUND ON INFORM'S FIELDS AND THEIR AVERAGE WATER-HOLDING CAPACITIES*

Soil name:	Percentages of three sizes of particles			Average water-holding capacity (in inches per foot of soil column)†
	Sand (2.0-.05mm)	Silt (.05-.002mm)	Clay (below .002mm)	
Yolo silty clay loam	25	50	25	2.2
Brentwood silty clay	10	55	35	2.2
Marvin silty clay loam	5	60	35	2.2
Capay silty clay loam	10	45	45	2.0
Panoche clay loam	35	40	25	1.9
Reiff fine sandy loam	30	55	15	1.8
Myers clay	15	45	40	1.8(pH>8)
Milham sandy loam	45	30	25	1.7
Nord fine sandy loam	45	40	15	1.6
Rincon silty clay loam	30	40	30	1.6
Garces loam	50	25	25	1.4(pH>8)
Grangeville fine sandy loam	15	40	45	1.3
Nahrub clay	15	40	45	1.3
Sacramento clay	5	40	55	1.2
Remnoy-Youd fine sandy loam	45	40	15	1.2
Kimberlina sandy loam	60	25	15	1.2(pH>8.4)
Traver fine sandy loam	55	30	15	1.2(pH>8.4)

* Source: U.S. Soil Conservation Service
† If water held dispersed in a foot-high column of soil (whatever the column's diameter) were gathered at the bottom of the column, it would occupy the number of inches indicated.

Soil Types

All the soil types found on INFORM's fields are products of alluvial deposits and thus have high proportions of silt and clay compared with the sandier soils of the California deserts to the south. Thus INFORM's sample was skewed toward fine-textured soils with relatively high water-holding capacities. Nonetheless, a wide range of soil types was encountered as indicated by Table 3 on the facing page.

Surface-Irrigation Methods

Border-strip irrigation systems were used on all 12 alfalfa fields in INFORM's study. On the four Sacramento Valley fields, irrigation water was drawn by siphons from open ditches. On the eight San Joaquin Valley fields, water was emitted through hydrants (alfalfa valves) tapping underground pipelines.

Furrow-irrigation systems were used on 19 fields of row crops (cotton, tomatoes, corn and sugar beets). On the nine furrow-irrigated Sacramento Valley fields (tomatoes, corn and sugar beets) open ditches and siphons were used for water applications. Four of the remaining 10 furrow-irrigated fields, all San Joaquin cotton, were irrigated with gated pipelines. The other six cotton fields were irrigated with hydrants on field layouts that combined furrows with the border-strip features typical of alfalfa fields. (See drawing in Appendix A, Field 12.)

The almond orchard analyzed in INFORM's 1984 pilot study was irrigated with sprinklers.

9
Using Gypsum-Block Data to Improve Water Distribution

In general, the more uniformly each irrigation refills a crop-root zone, the fewer irrigations are required during a growing season. When the root zone is evenly and fully watered, it sustains crop growth similarly on all field sections between irrigations, and the entire field need not be too frequently irrigated just to avoid yield loss on its underirrigated sections. Thus, identifying and reducing water-distribution problems are the essential first steps toward efficient irrigation.

INFORM learned how evenly farmers irrigated the 32 fields in its study by comparing gypsum-block readings from TOP, MID and END sites on INFORM's and the owners' strips. It used these findings to plan and conduct later experiments in reducing water applications on test strips on 22 of these fields. (See Appendix C for reasons why more fields were analyzed for water-distribution problems than were tested for water reductions.)

Summary of Findings

INFORM's gypsum-block data revealed that only seven of the 32 fields were evenly watered on both INFORM's and the owners' strips. One or both strips on the remaining 25 fields had one or more problems of uneven water distribution: Underirrigated TOP or TOP and MID sections at one or more depths (nine fields); underirrigated END or END and MID sections (nine fields); underirrigated depths throughout the root zone (three fields); continuous drying out of the field at most root-zone depths throughout the season (two fields). Finally, on eight fields the root zone was kept so wet throughout the season that distribution patterns were obscured.

(The total of these patterns exceeds 32 because in six cases different problems were discovered on INFORM's and the owners' strips during the same season, or on the same field when studied over two seasons.)

A Portrait of Uniform Distribution

The graph on page 72 shows the symmetrical patterns of uniform water distribution. These data come from a 1985 analysis of a Kern County alfalfa field where the farmer applied only one irrigation per monthly cutting cycle, compared to the two irrigations many farmers use, and produced a high 10 tons of hay per acre over the season's seven cuttings. (See Appendix A, Field 15.)

Three Examples of Detecting and Correcting Distribution Problems Using Soil-Moisture Data

INFORM tested changes intended to improve water distribution on 10 field strips in the 32 fields analyzed.

Part IV: INFORM's Field Work

Graph 3. An Evenly Watered Alfalfa-Field Strip

Each of the farmer's irrigations of this strip in an 80-acre Kern County alfalfa field studied in 1985 evenly refills the root zone at the three monitoring sites along a 1,250-foot run: Gypsum-block readings taken four days after the irrigations of July 12 and August 7 show meter values of 95 at all four depths at TOP, MID and END sites, indicating that the soil is uniformly saturated. Between irrigations (from June 17 - not shown - to July 12, and from July 12 to August 7), the similarly falling curves of moisture readings at TOP, MID and END sections along the run show the even drying out of the root zone.

Key		
	1 ft	■
irrigation }	2 ft	+
	3 ft	✶
cutting		4 ft	⊟
To read meter values, see page 38.		

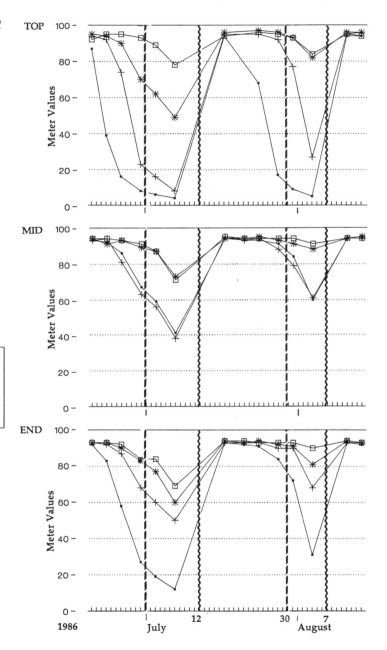

These 10 field strips were also among the 22 tested for water reductions.

The three following case studies illustrate INFORM's trial-and-error procedures for reducing distribution problems having common but very diverse causes: saline and alkaline conditions, an impermeable subsurface layer and a too-rapid rate of water application. (See Appendix B for two additional examples of uneven distribution.)

Case 1: Testing Sulfur Treatments to Open Up a Salt-Blighted Alfalfa Field

One of the nine fields revealed by INFORM's gypsum-block data to be underirrigated in TOP or TOP and MID sections in one or both strips, was an 80-acre alfalfa field in Kern County, that suffered from heavy clay soils, salt buildups and alkaline irrigation water. (Appendix A, Field 17.)

Although this field resembles many others in southern and western parts of the San Joaquin Valley, its problems were especially severe when INFORM began its 1985 analysis, says INFORM's cooperator, Randy Gafner. After each irrigation, water stood on the surface and its yields were the worst of any field he managed.

INFORM's 1985 gypsum-block data showed that water applied to the field's most blighted sections was penetrating only to the 1-foot depth and refilling far too small a portion of the root zone to sustain good crop growth. So, in 1986, Gafner and INFORM's technical consultant decided to use gypsum-block data in a trial-and-error experiment involving the addition of sulfur dioxide gas to irrigation water before irrigating.

In general, sulfur applications to treat saline soil conditions can often take two or more growing seasons

Randy Gafner of Kern County used gypsum blocks to add to what he "already knew" about his salt-blighted alfalfa field: that he wasn't getting water into the ground. He says that the blocks showed him just how severe the problem was, and gave him a reliable way to test one remedy for it by applying sulfurized water to his field.

Part IV: INFORM's Field Work

Graph 4. Sulfurized Irrigations Improve Water Penetration on an Alfalfa-Field Strip

Before. The May 8, 1986 irrigation of unsulfurized water has almost no effect on the severely salt-blighted TOP and MID sections of the owner's strip.

In the TOP section, the moisture levels are raised only slightly at the 1-foot depth, while no change is observed at the 2, 3 and 4-foot depths. In the MID section, the 1, 2 and 3-foot depths are unaffected. At the 4-foot depths in both TOP and MID sections, odd rises in moisture levels occurring several days after irrigations cannot be explained without soil analysis.

After. The August 16 irrigation, the sixth to have been treated with sulfur dioxide gas, penetrates and replenishes the 1-foot depth in the TOP section and the 1 and 2-foot depths in the MID section. Moreover, the water-storage capacity of the MID section's 1-foot depth is greatly improved: Moisture levels remain at saturation for 10 days after the irrigation. The skipping of the 2-foot depth on the END section occurred because not enough water reached the end of the field.

The 10-day gap in readings between July 28 and August 7 probably obscures a rise and fall

continued

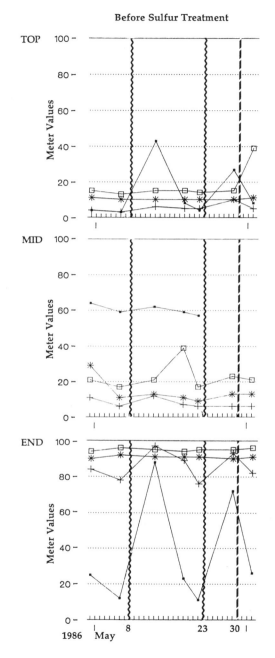

Using Gypsum-Block Data to Improve Water Distribution

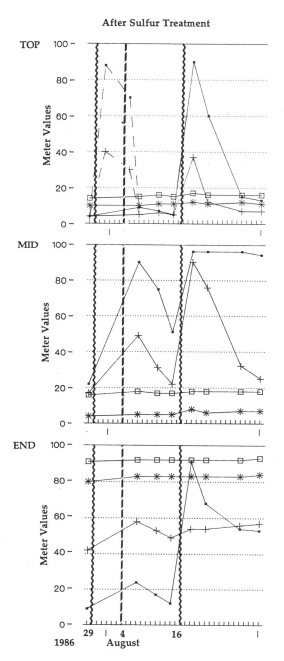

On Randy Gafner's "worst" field, his alfalfa wouldn't grow at all in some sections (lighter areas occupying most of photo) because of salt buildups, heavy clay soils and alkaline irrigation water.

Graph 4, continued

in moisture levels in the TOP section's 1 and 2-foot depths after the irrigation of July 29. The dashed curves indicate the likely pattern of moisture changes during these 10 days.

Key		
	1 ft	-■-
irrigation ⎫	2 ft	+
	3 ft	✳
cutting ⎪	4 ft	⊟
To read meter values, see page 38.		

Part IV: INFORM's Field Work

Bill Longfellow says that alfalfa hay from a less-irrigated field strip is higher in nutrients than hay from a more-heavily irrigated strip. Longfellow and his brother want to boost the nutrient content of the alfalfa they grow to feed their dairy cows because better hay produces better milk.

to produce yield improvements. Moreover, only certain types of salinity problems are treatable, and these are costly to analyze precisely using complex laboratory procedures. However, as Gafner found and the graphs on pages 74-75 reveal, soil-moisture data can give an inexpensive trial-and-error answer within one season.

Case 2: Finding an Impermeable Soil Layer on a Releveled Alfalfa Field

One of the nine fields shown by INFORM's data to be underirrigated in END or END and MID sections of one or both strips, was a 40-acre alfalfa field in Kings County. (Appendix A, Field 19.) Here a recent releveling had relayered soil types and created an impermeable layer at the 2-foot depth on the END section of INFORM's strip.

In its 1986 test, INFORM first ruled out insufficient water application as the cause of the problem by opening the hydrants on its strip more fully and leaving them open longer for the July 6 irrigation. The heavier watering flooded the END section of the INFORM strip and even spilled into adjacent checks through flat spots in the machine-damaged berms. Yet, block data gathered after the inundation showed that it had had no impact on the penetration problem. Soil at the 3 and 4-foot depths was as dry as before the irrigation, and soil at the 2-foot depth had been only partly rewetted, as the graph on facing page shows.

Intrigued by the test results, INFORM's cooperator, Bill Longfellow, ripped the END section of his field in the fall of 1986. "As soon as I got down to that bottom end, I sure could tell," he says. "The soil was so tight at about two feet that I had to get off the tractor and pull the ripper out of the ground by hand."

Using Gypsum-Block Data to Improve Water Distribution

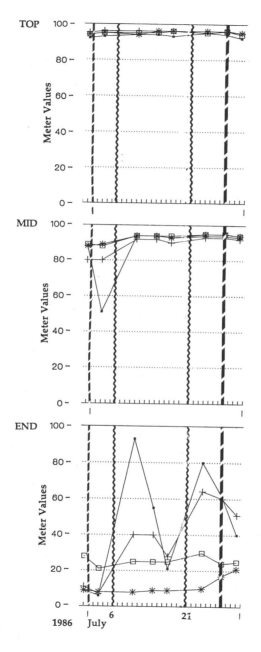

Graph 5. An Impermeable Layer on an Alfalfa-Field Strip

After learning from July 10, 1986 soil-moisture readings that a July 6 flooding did not increase water penetration below the 2-foot depth on this field strip's END section, cooperator Bill Longfellow ripped the troublesome section at the end of the season. He broke up the hard soil layer created during a recent releveling, thereby removing the cause of the problem.

Key		
	1 ft	-■-
irrigation ⌇	2 ft	+
	3 ft	✳
cutting ┆	4 ft	⊟
To read meter values, see page 38.		

Part IV: INFORM's Field Work

Leroy Bertolero of Solano County illustrates in the dust how he irrigates his short 500-foot runs that help him manage water efficiently on 1,000 acres of processing tomatoes. Bertolero says that INFORM's tests on one tomato bed gave him the evidence he needs to further improve his practices on a number of fields, thereby reducing his pumping costs and protecting ripening fruit from rot by keeping the field surface drier.

Case 3: Evening Out the Wetting Patterns Beneath a Tomato Bed

One of the three fields shown by INFORM's data to have consistenly underwatered depths on one or both strips was a 21-acre tomato field in Solano County. Here the farmer's irrigations skipped the shallow sections beneath the center of a tomato bed's TOP and END sections on a short, 500-foot run. (Appendix A, Field 3.) INFORM modified its usual pattern of gypsum-block monitoring, as illustrated on the facing page, to detect this problem and test slower and longer irrigations to solve it. These adjustments, whose results for the TOP section are shown on the graph on page 80, nearly eliminated the distribution problem.

They also had other benefits, according to INFORM's cooperator, Leroy Bertolero. The longer and slower final irrigation used 20% less water on the test bed than on other beds in the field. It did not wet the bed's surface where it could have rotted low-hanging tomatoes in contact with the soil. And "not a drop" of water ran off into the drain ditch. "It worked out so well we did the same thing on some of the other fields that we planted later that year," Bertolero reports.

Better Distribution as a Condition of More Efficient Scheduling

INFORM's 1984 test on one alfalfa-field strip in an 80-acre field in Yolo County, illustrates clearly how improvements in water distribution can lead to the use of fewer irrigations. (Appendix A, Field 4.) On this field, the farmer applied each irrigation too rapidly to allow it sufficient time to refill the owner strip's TOP section. Then, whenever the soil surface dried out on the TOP

Using Gypsum-Block Data to Improve Water Distribution

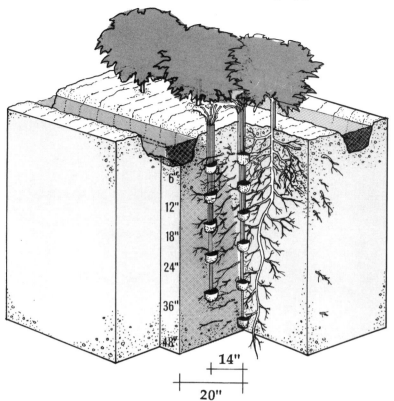

Dry Center-Bed Sections on a Tomato-Field Strip

Water percolating into a tomato bed from furrows on either side did not fully rewet the upper soil depths beneath the bed's center. INFORM detected and corrected this problem using a special monitoring plan. A pair of stations was placed in the bed's TOP section 100 feet from the supply ditch, and another pair in its END section 100 feet from the drain ditch. Both center-bed stations monitored 6, 12, 18, 24, 36 and 48-inch depths among the deepest growing plant roots. The furrow-edge stations where root growth was shallower did not include a 48-inch-deep block.

Part IV: INFORM's Field Work

Graph 6. Using Slower Irrigations to Wet the Center Section of a Tomato Bed

INFORM's data showed that the first three irrigations of the 1986 growing season applied on June 9, June 23 and July 2 failed to refill the shallower depths under the TOP center section of this tested tomato-bed strip. All three irrigations were applied for 12 hours, but the first two used 1-1/2-inch siphons, one per furrow, and the third used 1-inch siphons. The problem can be seen by comparing the TOP-center station's post-irrigation readings of June 11, June 27 and July 5 with readings taken on the same dates from the TOP-furrow-edge station.

When INFORM lengthened the July 12 irrigation to 24 hours, using 1-inch siphons, the moisture levels at the 12-inch depth in the bed's TOP-center section were replenished for the first time in the season. Then, on July 24, when INFORM substituted 3/4-inch siphons and further lengthened the irrigation to 32 hours, the moisture levels were restored at all depths but the 6-inch depth where the impact of the skipping was likely to be very slight. These same adjustments also corrected a similar problem in the bed's END-center section.

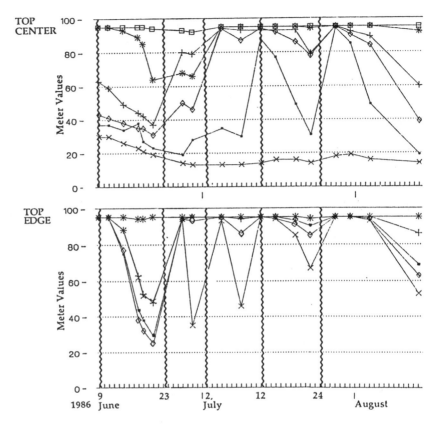

Key		irrigation ⌇	
6 in.	×	24 in.	+
12 in.	■	36 in.	✳
18 in.	◇	48 in.	⊟

To read meter values, see page 38.

Using Gypsum-Block Data to Improve Water Distribution

Bill Roth of Yolo County (left), tells Peter Mueller-Beilschmidt that he's "got the principle down": By running water onto his alfalfa fields more slowly and for more hours, he gets water more evenly into the ground and uses fewer irrigations per season.

section, the farmer irrigated, even though other field sections were still very wet. This practice caused him to apply nearly twice as many irrigations as the INFORM strip needed once the distribution problem on it was solved by using slower and longer irrigations. Instead of using 12-hour irrigations with four 4-inch siphons per check applied by the farmer on the owner's strip, INFORM corrected the distribution problem on its strip by removing two of the four siphons after 6 hours and running the water through the remaining two siphons for another 18 hours. (See Graph 7.)

Graph 7. Improved Distribution Reduces Irrigation Frequency

In 1984, water applied too rapidly on this alfalfa field never refilled the TOP section of the owner's strip. INFORM corrected the problem on its strip by using slower and longer irrigations. Over four months, including the three shown here, INFORM used 3 acre-inches more water per acre for each irrigation, but applied only four irrigations compared to the seven applied to the owner's strip.

Key	
irrigation ⎰	1 ft •-
	2 ft +
cutting ⎱	3 ft ✳
	4 ft ⊟
To read meter values, see page 38.	

(overleaf)

Part IV: INFORM's Field Work

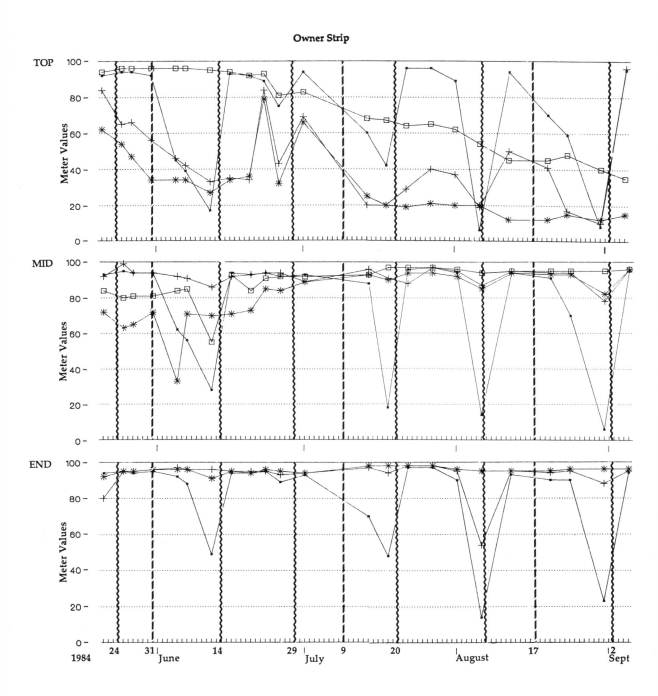

Using Gypsum-Block Data to Improve Water Distribution

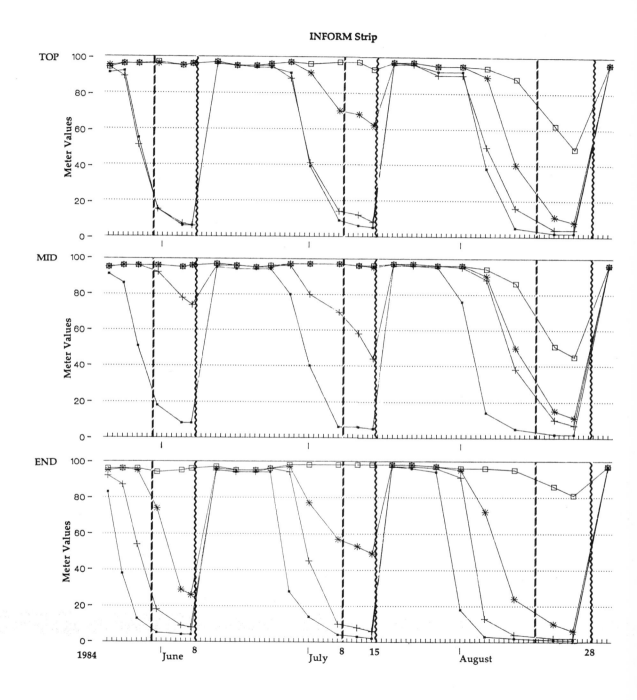

10
Using Gypsum-Block Data to Schedule Irrigations and Reduce Water Use

From 1984 to 1986, INFORM used soil-moisture data to reduce water use on 22 field strips, one of which was tested for two seasons. Changes introduced to achieve these water reductions included using smaller and/or fewer siphons on four strips, fewer irrigations on 14, and both adjustments on four strips.

Three Examples of Irrigation Scheduling

INFORM's typical trial-and-error procedures for using soil-moisture data to test lighter water applications and less frequent waterings are illustrated by the graphs from three fields presented on pages 86-91. All three fields were being evenly and fully wetted to the 4-foot depth by the farmers' water applications, and all three showed the same or higher yields on INFORM's less-irrigated strips. The three examples include: using smaller siphons on a tomato strip (Graph 8); halving the number of summer irrigations on an alfalfa strip (Graph 9); and eliminating three of seven irrigations on a cotton strip (Graph 10).

The examples of the alfalfa and cotton fields show how INFORM used block data from a "critical" depth for scheduling irrigations. On alfalfa fields, this depth was 3 feet because deep-rooted alfalfa plants normally draw about half of their water supply from the soil's top 3 feet. On cotton fields, it was 2 feet, because the shallower-rooted cotton plants normally draw about half of their water from the top 2 feet of soil. In both cases, the aim was to schedule irrigations at about the time the 50% depletion level was reached as indicated by rapidly falling or low moisture levels at the critical depth. Letting root zones dry out to this extent between waterings is widely recognized as being linked with good yields for both crops.

Bracketing Procedures and a Reduced Pre-Irrigation

INFORM demonstrated, on one field strip each, two additional ways to use gypsum-block data to schedule irrigations efficiently: by bracketing the minimum water application needed on an alfalfa field strip (see Graph 11, page 92) and by reducing the amount of a pre-irrigation on a cotton field. These methods would have been demonstrated on more fields if INFORM had tested its fields for a second season.

Bracketing the Minimum Water Application on an Alfalfa Field Strip. By trial-and-error tests over several irrigations, a farmer can gradually reduce the length and/or volume of irrigations until eventually he applies one irrigation that fails to fill the root zone fully. When this happens the minimum irrigation length and rate (and thus the unmeasured volume) required on that field strip has been "bracketed." This length and rate are greater than those of the insufficient irrigation but less than those of the one preceding it.

Part IV: INFORM's Field Work

Graph 8. Gypsum-Block Data Show that Smaller Siphons Safely Reduce Water Use on a Tomato-Field Strip

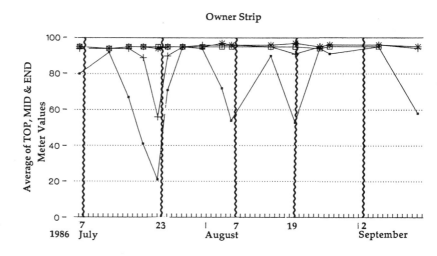

By using smaller siphons and eliminating one irrigation, INFORM used only 10.4 acre-inches of water per acre on its test strip, compared to 24 acre-inches used on the owner's strip, in a 35-acre Yolo County tomato field tested in 1986. (Appendix A, Field 1.) INFORM used soil-moisture data from both strips to guide and evaluate its testing as follows:

1) The readings of July 5 on INFORM's strip showed soil saturation to the 3-foot depth. Even so, INFORM applied the unneeded irrigation of July 6 to try out one 1-inch siphon per furrow for 12 hours, instead of the one 2-inch siphon per furrow used for 6 hours by the farmer. Each 1-inch siphon pulled 9 gallons per minute (gpm) from the ditch onto the field compared with the 35 gpm delivered by each 2-inch siphon. (See Appendix D on siphons.)

2) The smaller July 6 irrigation streams on the INFORM strip dried up before reaching the end of the run, but this did not deprive the already wet root zone of needed supply. To avoid the drying up of the small streams during the July 23 irrigation, INFORM used two 1-inch siphons (18 gpm per furrow) for 6 hours, until the water streams reached the drain ditch, and then removed one of the siphons and left the other in place for another 6 hours.

3) INFORM repeated the July 23 procedures on August 7 and September 3, and omitted the irrigation applied to the owner's strip on August 19 because readings on August 14 and 19 showed high moisture levels on its strip. Block readings in August showed a slower drying out of INFORM's than the owner's strip and indicated that slower, lighter and longer irrigations had stored more water in the root zone than the owner's faster, heavier and shorter irrigations.

Using Gypsum-Block Data to Schedule Irrigations and Reduce Water Use

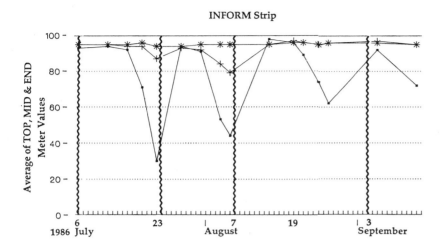

Bryan Barrios of Yolo County talks from a tomato harvester's cab about hard times and the need to cut back costs. But he also says there are a lot of rewards in farming.

Part IV: INFORM's Field Work

Graph 9. Monitoring the 3-Foot Depth on an Alfalfa-Field Strip to Apply One Instead of Two Irrigations per Growth Cycle

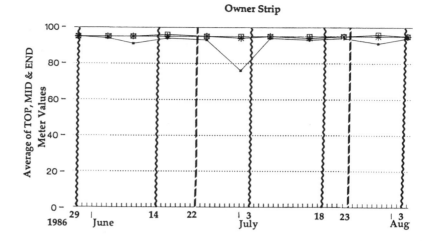

By monitoring the 3-foot depth, INFORM scheduled one irrigation per monthly growth cycle on its test strip, compared with the two irrigations on the owner's strip, in an 80-acre Yolo County alfalfa field tested in 1986. (Appendix A, Field 7.) June and July readings from INFORM's and the owner's strips show how INFORM used soil-moisture data to achieve these results.

1) June and July readings from the owner's strip showed that moisture levels at all four monitored depths remained continuously at or above field capacity (about 90 on the meter scale) except for a slight dip at the 1-foot depth at the end of July. Thus INFORM had reason to believe that the mid-growth-cycle irrigations of June 14 and July 18 were unnecessary.

2) INFORM tested and confirmed this hypothesis by eliminating the June 14 and July 18 irrigations on its test strip and noting that the 3-foot moisture levels nonetheless remained high throughout the two-month period.

In addition, INFORM applied less water with each of its fewer applications by using fewer siphons on its strip. Overall, INFORM's strip used 2.8 acre-feet of water per acre compared to 6.7 acre-feet per acre used on the owner's strip.

Key		
		1 ft -■-
irrigation }		2 ft +
		3 ft ✻
cutting ∤		4 ft ⊟
To read meter values, see page 38.		

Using Gypsum-Block Data to Schedule Irrigations and Reduce Water Use

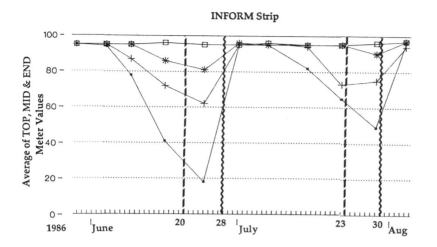

Chester Roth of Yolo County figures he can find something else to do with the $4,000 per year that INFORM's test indicated he was losing by overwatering an 80-acre alfalfa field in 1986.

Part IV: INFORM's Field Work

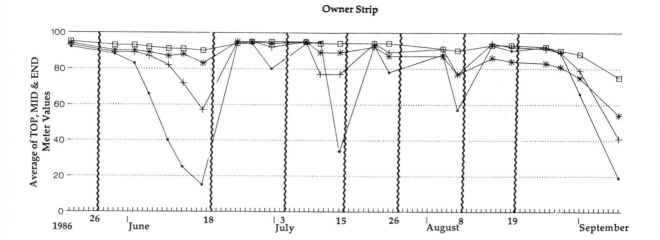

Graph 10. Monitoring the 2-Foot Depth on a Cotton-Field Strip to Eliminate Three Irrigations

By irrigating only when moisture levels were dropping at the 2-foot depth, INFORM used three fewer irrigations on its strip than were used on the owner's strip during the 1986 season on an 80-acre Kern County cotton field. (Appendix A, Field 24.) INFORM's strip used 32 acre-inches per acre for the season compared with the 34.5 acre-inches per acre used by the owner's strip.

In 1986 the farmer had already reduced water use on the owner's strip compared with 1985, by using expensive, high-tech surge-irrigation equipment that emitted water into the furrows in spurts. By contrast, INFORM's strip was irrigated with a standard gated pipeline. Yet it used even less water than the owner's strip when INFORM's scheduling decisions were guided by soil-moisture data as follows:

1) Late May readings from both strips showed that soil at all four depths was still saturated four weeks after the planting date. This indicated that the young cotton roots had not yet depleted the water supply even in the shallow sections of the root zone and thus that they were still small. In order to give the roots more time to grow into lower soil levels (to develop strong root systems), INFORM asked that the May 26 irrigation be omitted from its strip.

2) At the time of the mid-June irrigations, falling moisture levels at the 2-foot depth on both strips showed that crop roots had penetrated to this depth. Even these irrigations could have been delayed since substantial depletion at this depth had not yet occurred.

3) The July 6 irrigation on INFORM's strip was applied at a time of rapidly falling moisture levels at the 2-foot depth caused by the crop's high water demand during the flowering and setting of cotton bolls. During such reproductive periods delayed irrigations may damage yields. However, the July 3 irrigation on the owner's strip was applied before the 2-foot moisture levels had begun to drop.

4) The July 19 irrigation on INFORM's strip was not needed but was nonetheless applied because it was difficult for the farmer to return to irrigate this one strip separately at a later date. (In general, INFORM's

Using Gypsum-Block Data to Schedule Irrigations and Reduce Water Use

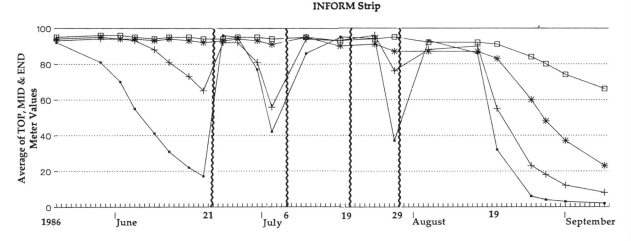

INFORM Strip

irrigation suggestions could be followed only when cooperators could arrange to treat the tested field strips differently from the rest of the crop.) The July 26 irrigation was applied earlier than necessary on the owner's strip.

5) In late July, falling moisture levels at the 2-foot depth on INFORM's strip led INFORM to apply the July 29 irrigation. However, the farmer's August 8 irrigation was omitted on the still wet INFORM strip. The farmer's August 19 irrigation was also omitted on INFORM's strip due to an error in communication. This mistake resulted in a faster drying out of the INFORM strip at the season's end, but the farmer reported no observable difference in yields.

Key — irrigation:
- 1 ft ■
- 2 ft +
- 3 ft ✶
- 4 ft ▤

To read meter values, see page 38.

Gary Wilson of Kern County is always searching for ways to cut back on irrigation water costing him $40 an acre-foot to pump. INFORM's gypsum-block tests indicated that, by spacing irrigations farther apart, he could eliminate one or two irrigations of the five to seven irrigations he applies per season.

Part IV: INFORM's Field Work

Graph 11. Bracketing Procedures on an Alfalfa Field Strip

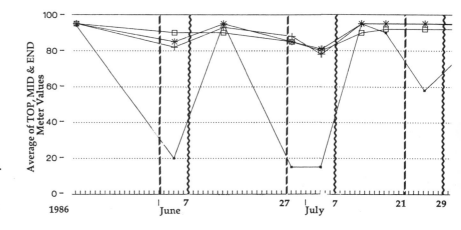

INFORM's strip on a 38-acre alfalfa field in Yolo County was watered on June 7 and July 7, 1986 through leveled spots (cutouts) in the supply ditch's bordering levees. Pump data indicated that 8.7 acre-inches of water were applied on each date. INFORM asked that the ditch be repaired. Then five 3-inch siphons per check were used for the July 29 irrigation. When block data showed that this lighter application of 6.8 acre-inches per acre succeeded in refilling the root zone, INFORM tried the August 29 irrigation with only four 3-inch siphons per check, and applied only 4.9 acre-inches per acre. However, because the August 29 irrigation failed to refill the crop root zone completely, INFORM deduced that the minimum application needed was less than 6.8 acre-inches per acre but more than 4.9 acre-inches per acre.

Key		
	1 ft	-■-
irrigation }	2 ft	+
	3 ft	✶
cutting │	4 ft	☐
To read meter values, see page 38.		

On a 38-acre alfalfa field in Yolo County, data gathered after two heavy waterings of INFORM's strip and data gathered after two later, lighter irrigations, bracketed the minimum application needed. (See graph above and Appendix A, Field 2.) Overall, INFORM's strip used 29 acre-inches per acre and the owner's strip used 34 acre-inches per acre. INFORM's yields were slightly higher.

The one requirement for using such bracketing methods is that the crop-root zone be similarly depleted at the time of each irrigation in the step-by-step sequence of water reductions. This requirement was closely approached in INFORM's tests.

Reducing Pre-Irrigation on a Cotton Field. Farmers often irrigate cotton fields before planting to ensure that the soil-moisture levels are at or near field capacity. These wet conditions are essential for the timely germination of seeds and for the rapid growth of new root systems. (Rootlets cannot penetrate dry soil.)

Gypsum-block data collected before such pre-irrigations can be used to estimate how much water must be

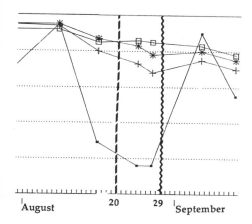

added to the moisture already restored by winter rain to bring the field to capacity. Readings taken after the pre-irrigation show if field capacity is reached or exceeded.

INFORM monitored pre-irrigations for two seasons on one 80-acre cotton field in Kern County. (Appendix A, Field 24.) In 1985, INFORM estimated that 3 acre-inches of water per acre were needed to refill the root zone's top 4 feet, but the farmer applied 14 acre-inches per acre. In 1986, INFORM estimated that 5 acre-inches per acre of irrigation were needed on the field and asked that this amount be applied to its 0.4-acre strip. By contrast, the farmer applied 11 acre-inches per acre to the rest of the field. His irrigation "lost" about 6 acre-inches per acre, or about 40 acre-feet for the entire field, to deep percolation below the root zone's top 4 feet.

In testing pre-irrigations, estimates of the irrigation volumes needed to refill to capacity different soil types at different levels of moisture depletion can be acquired from various sources by farmers. Then, trial-and-error

Part IV: INFORM's Field Work

field tests show farmers how to link gypsum-block readings with particular soils' moisture-depletion levels in order to schedule pre-irrigations.

Part V
Benefits Achieved

11
Water Reductions and Yield Increases

Between 1984 and 1986, INFORM successfully conducted 21 of its 23 tests of water reductions on INFORM-managed strips. Two of INFORM's 23 tests were invalidated. In one case, the farmer applied a pre-irrigation to the owner's strip but not to INFORM's, thus giving a critical initial advantage to the owner-strip crop. In the second, the farmer's water applications were so irregular that INFORM's strip actually received more water than the owner's strip despite its two fewer irrigations.

INFORM's results show the efficacy of the soil-moisture method for improving irrigation practices.

Water Use Reduced Up to 58%

In 19 of the 21 tests, INFORM's strips used 6% to 58% less water than the owners' strips. On two other fields belonging to the same farmer, INFORM applied fewer irrigations to the test strips, but could not estimate the

Part V: Benefits Achieved

Wayne Wisecarver of Kings County doesn't believe any promises until he sees field proof of results. He says that soil-moisture monitoring will soon be a necessity for most farmers in the southern San Joaquin Valley where water and energy costs have tripled since the mid 1970's.

percentages by which water use was thus reduced because the lengths and volumes of the farmer's irrigations were irregular.

The volumes of water reduction on the 15 INFORM strips where volumes could be estimated ranged from 0.25 acre-feet (81,463 gallons) per acre to 3.9 acre-feet (1,270,819 gallons) per acre.

INFORM estimated the *percentages* by which applications on 19 strips were reduced by relying on farmers' judgments about the "sameness" of irrigations. When a farmer reported that each irrigation was about equal in application rate and was applied for about the same length of time, INFORM considered each irrigation an equivalent percentage of the total volume of waterings.

INFORM estimated the *volumes* of water reductions on strips in the 15 fields irrigated with siphons or gated pipelines because the outflow rates from this equipment could be reasonably and practically estimated. However, INFORM did not estimate the volumes of water reductions on six strips in fields irrigated with hydrants, including the two irregularly irrigated fields, because no practical, field-level method exists for making such estimates. In general, INFORM used only tools and methods that farmers themselves can easily acquire and use.

Crop Yields Improved on 10 Fields

Yields on INFORM-managed strips were higher than on owners' strips in 10 cases, the "same" in six cases, impossible to compare (according to the farmer) in one case, and lower in four cases. Two fields produced lower yields because soil conditions on INFORM's strips were so poor that water reductions could be achieved only with some yield loss. However, none of the higher yields on INFORM's strips were caused by noticeably better soil conditions than those found on the owner's strips. On the two other low-yield fields there were communication problems that caused delays in early-season irrigations. These delays rather than the volume of reductions were the cause of yield losses, according to the farmers.

Yields were measured or estimated in three ways. On the 10 alfalfa fields, INFORM counted the bales of hay produced by INFORM's and owners' strips. On the 11 fields of row crops, farmers measured yields in four cases and estimated yields in the remaining seven.

Table 4 on pages 100-101 presents the findings of INFORM's 21 valid field tests. The table's footnotes describe the site-specific considerations that were most relevant to evaluating test procedures and results. Additional site-specific details are provided for each field in Appendix A.

Yield increases on seven alfalfa-field strips where INFORM tested water reductions ranged from 80 to to 1,000 pounds per acre per cutting. Alfalfa fields are normally cut five to seven times a season.

Untested Opportunities for Further Reductions

On every field, INFORM identified opportunities for achieving larger water reductions than actually were demonstrated. Time was the principal constraint since during one growing season only a certain number of changes could be tried. In general, several growing seasons are needed for farmers to convert to new practices of irrigation management because these changes must be tested during and after irrigations. They must also be integrated with many other parts of the farm operation, from fertilization programs to the deployment of pumps and labor.

Specific changes in water management that, if tested over one or more additional seasons, would likely have

Part V: Benefits Achieved

TABLE 4. WATER REDUCTIONS AND CROP YIELDS IN 21 TESTS

Notes to table:

1. One acre-foot equals 325,851 gallons. Volume estimates, where given, were based on measured outflow through siphons (S) or through gated pipelines (G), or on pump data (P).

2. Based on INFORM's bale counts for alfalfa (B), or on farmers' measurements (M) or reports (R).

3. Field observations and pump data from this field, combined with the one acre-foot per acre of water reductions easily achieved during tests in August and September, indicated that the farmer was applying more than twice as much water as the field needed.

4. Because of the quality of the field data and the farmer's testimony, results for part of the season are extrapolated to the season as a whole.

5. The owner imitated INFORM's irrigation changes, and claimed that 20% less water per acre was applied to both strips than to nearby fields. Yield losses on INFORM's strip, measured at 11% by the farmer, were attributed by him to poorer soil conditions on INFORM's side; yields on the owner's strip, which was watered identically to INFORM's, were comparable to those on other nearby fields.

6. This field belonged to a farmer who participated in INFORM's 1984 pilot study and who was already using the slower application rates shown in that study to reduce his water use. Thus, in 1985, this field was irrigated with 1-inch siphons, and results on it were compared by

Field (See Appendix A)	Percent of water reduction on INFORM strip	Per-acre volume of water reduction on INFORM strip (acre-feet)[1]	Yields on INFORM strip[2]
I. Fields where changes on INFORM's strip included slowing down the water-application rate and reducing the number of irrigations			
#4 Alfalfa (80ac) Yolo County-84	14	0.50 (S)	higher(B)
#28 Alfalfa (100ac) Yolo County-85	50[3]	1.00 (S)	slightly higher(B)
#7 Alfalfa (80ac) Yolo County-86	58	3.90 (S)	slightly higher(B)
#1 Tomatoes (35ac) Yolo County-86	57	1.13 (S)	same (R)
Averages	**45**	**1.6**	
II. Fields where changes on INFORM's strip involved slowing down the water-application rate only			
#2 Alfalfa (38ac)[4] Yolo County-86	29	0.83 (S)	slightly higher(B)
#11 Tomatoes (35ac)[5] Solano County-84	20	0.41 (S)	same (M)
#11 Tomatoes (35ac)[6] Solano County-85	19	0.35 (S)	very high yields and high sugar content (6.2% solids) of tomatoes for whole field(R)
#3 Tomatoes (21ac) Solano County-86	34	1.00 (S)	same(R)
Averages	**26**	**0.65**	

100

Water Reductions and Yield Increases

Field (See Appendix A)	Percent of water reduction on INFORM strip	Per-acre volume of water reduction on INFORM strip (acre-feet)[1]	Yields on INFORM strip[2]
III. Fields where INFORM reduced the number of irrigations on its strip but did not change the rate of water application			
#13 Alfalfa (160ac,No.) Tulare County-86	29[7]	NA	substantially higher(B)
#14 Alfalfa (160ac,So.) Tulare County-86	29[7]	NA	substantially higher(B)
#23 Alfalfa (27ac) Fresno County-86	22	0.63 (P)	higher(B)
#19 Alfalfa (40ac) Kings County-86	NA[8]	NA	higher(B)
#25 Alfalfa (80ac)[9] Kings County-86	14[7]	NA	lower(B)
#17 Alfalfa (80ac)[9] Kern County-86	14	0.58 (G)	lower(B)
#8 Corn (30ac) Solano County-84	27	0.50 (S)	lower(M)[10]
#10 Corn (50ac) Solano County-85	6	0.25 (S)	same(R)
#24 Cotton (80ac) Kern County-85	28	1.17 (G)	lower(M)[10]
#24 Cotton (80ac) Kern County-86	18	0.67 (G)[11]	same(R)
#16 Cotton (80ac) Kern County-86	29	1.17 (G)	higher(M)[12]
#27 Cotton (52ac) Kings County-86	20[7]	NA	NA(R)[13]
#21 Cotton (40ac) Kings County-86	NA[8]	NA	same(R)[14]
Averages (III)	**21** (11 strips)	**0.71** (7 strips)	

the farmer with results on nearby fields that were irrigated with larger siphons.

7. The estimate of the percentage of water reductions on INFORM's strip is based on the difference in the number of irrigations applied to the owner's and INFORM's strips.

8. Each irrigation was applied at a different rate, for a different length of time. On Field 19, two fewer irrigations were applied to INFORM's and the owner's strips than to the rest of the field; on Field 21, one fewer irrigation was applied to INFORM's strip.

9. The INFORM strip on this field was tested for only one growth cycle because severe salt problems impeded water penetration.

10. A delayed irrigation in June rather than the reduced water use per se, probably lowered yields, according to the farmer.

11. This also included a reduction in the pre-irrigation.

12. See Appendix A, Field 16, for the farmer's specially devised test of yield differences.

13. The farmer said the two sides of the field had such different soil types that a comparison of yields made no sense.

14. Here the farmer was unsure about the effect of water reductions on yields, but INFORM's data showed such a slight drop in root-zone moisture levels caused by applying one fewer irrigation to INFORM's strip, that yields were judged the same.

Part V: Benefits Achieved

increased the size of the reductions on INFORM's strips include the following:

• On all field strips tested, bracketing procedures could have been used to ascertain the minimum amount of water to apply with each irrigation. (See Chapter 10.)

• Improvements in water distribution could have been initiated or extended on at least 10 field strips.

• Fewer irrigations per season might have been applied on eight to 10 test strips in fields of row crops.

• Pre-planting irrigations could have been monitored and probably reduced on six cotton-field strips.

• On three alfalfa fields, a one-irrigation-per-cutting schedule could have been extended from one, two or three cuttings to all cuttings of the growing season.

12
Benefits and Costs of the Soil-Moisture Method

INFORM demonstrated financial benefits in 14 of its 21 valid field tests with gypsum blocks. The benefits for one growing season, on the tested strips, ranged from less than $1 per acre for water-cost savings alone, to over $165 per acre for both water-cost savings and the value of higher yields.

No dollar benefits were identified for the remaining seven of the 21 tests. In four cases, yield losses occurred and water reduction benefits were not measured. In three cases, including one where water use was lower and yields higher on INFORM's strip, neither water use nor yields could be reliably estimated.

Water-Cost Savings

Water-cost savings for one season of $1 to $39 per acre were achieved in 12 of the 16 INFORM tests where yields on INFORM's strip improved or remained the same. These savings came from the reduced costs of buying or pumping water. (See Table 5.) On four of the 16 field

Part V: Benefits Achieved

TABLE 5. WATER-COST SAVINGS ON INFORM'S FIELD STRIPS WHERE YIELDS WERE HIGHER OR THE SAME

Field (See Appendix A)	Water cost per acre-foot	Size of tested INFORM strip (acres)	Water source	Reductions on INFORM strip in acre-feet	Savings per acre on INFORM strip
#1 Tomatoes(35ac) Yolo County-86	< $1.00	0.25	Drain Ditch	1.1	< $1.00
#10 Corn(50ac) Solano County-85	$17.00	0.62	Well	0.25	$4.25
#4 Alfalfa(80ac) Yolo County-84	$10.00	5.40	District	0.5	$5.00
#11 Tomatoes(53ac) Solano County-85	$17.00	0.65	Well	0.35	$5.95
#9 Tomatoes(35ac) Solano County-84	$17.00	17.00	Well	0.41	$6.97
#23 Alfalfa(27ac) Fresno County-86	$14.00	1.70	Well	0.63	$8.82
#2 Alfalfa(38ac) Yolo County-86	$12.00	3.00	Well	0.83*	$9.96
#3 Tomatoes(21ac) Solano County-86	$17.00	1 tomato bed	Well	1.0*	$17.00
#28 Alfalfa(100ac) Yolo County-85	$22.50	3.70	Well	1.0	$22.50
#16 Cotton(80ac) Kern County-86	$33.00	0.50	Well	1.17	$38.61
#7 Alfalfa(80ac) Yolo County-86	$10.00	2.90	District	3.9	$39.00
#24 Cotton(80ac)† Kern County a) 1985 (lower yields) b) 1986 c) INFORM 1986 compared with farmer 1985	$40.00 $40.00 $40.00	.40 .40 .40	Well Well Well	1.17 .67 2.25	$46.80 $26.80 $90.00

104

strips showing equivalent or higher yields, no estimates of water reductions or water savings could be made because no practical means was available for measuring water outflow through the hydrants used for irrigating.

Savings approaching $90 per acre over two growing seasons were achieved on the one INFORM field strip, in a cotton field, that was tested in both 1985 and 1986. (See footnote to Table 5.)

The Value of Yield Increases

On eight of the 10 INFORM test strips that produced higher yield quantities, the values of these yield increases ranged from $5.60 to $126.50 per acre. (See Table 6.) On the remaining two field strips, farmers reported better crop quality but no reliable dollar values could be assigned to the improvements: On one alfalfa field, the farmer sent hay from the drier INFORM strip to a laboratory for testing and found that it was higher in nutrient content than hay from a nearby wetter field section. On one tomato field, the farmer reported very high sugar content (6.2% solids) for tomatoes that had been watered 20% less on both INFORM's and the owner's strips than on nearby fields.

Craig Fulwyler, who manages McConnell Farms in Kern County, devised a way of measuring 1986 cotton yields on and near INFORM's field strip to learn that two fewer irrigations produced an additional 0.41 bales of cotton per acre, worth over $126 per acre in increased earnings.

---

* Season-long estimates based on tests over two irrigations and confirmed by the farmer.

† Tests were conducted on this field for two years, and a comparison of both seasons is required to evaluate the savings in water and cost. In 1985, the farmer applied 64 acre-inches per acre to the field including 14 acre-inches applied during pre-irrigation. During the next season, INFORM applied 37 acre-inches per acre to its field strip including 5 acre-inches for pre-irrigation, whereas the farmer applied 45 acre-inches per acre including 10.5 acre-inches for pre-irrigation. The total water-reductions demonstrated on this field—comparing the farmer's application in 1985 with INFORM's in 1986—was thus 27 acre-inches or 2.25 acre-feet per acre. The savings in water costs on the INFORM strip over two seasons was about $90 per acre.

Yields in 1985 were lower on INFORM's strip due to a delayed early irrigation. However, INFORM's yields in 1986 were judged by the farmer to be the "same" as on his strip, even though 13 fewer acre-inches of water per acre were applied to the INFORM strip in 1986 than in 1985. This evidence leads INFORM to conclude that its two tests demonstrated a water-cost savings approaching $90 per acre, without yield loss.

Part V: Benefits Achieved

TABLE 6. VALUE OF MEASURED YIELD INCREASES ON INFORM'S FIELD STRIPS

Field (See Appendix A)	Size of INFORM strip (acres)	Increased yield on INFORM strip (per acre)*	Increased revenues on INFORM strip (per acre)†
#28 Alfalfa(100ac) Yolo County-85	3.7	0.07 tons (one cutting)	$5.60
#7 Alfalfa(80ac) Yolo County-86	2.9	0.2 tons (four cuttings)	$16.00
#2 Alfalfa(38ac) Yolo County-86	3.0	0.2 tons (four cuttings)	$16.00
#4 Alfalfa(80ac) Yolo County-84	5.4	0.25 tons (three cuttings)	$20.00
#13 Alfalfa(160acN) Tulare County-86	12.4	1.05 tons (two cuttings)	$84.00
#23 Alfalfa(27ac) Fresno County-86	1.7	1.1 tons (four cuttings)	$88.00
#14 Alfalfa(160acS) Tulare County-86	12.4	1.55 tons (three cuttings)	$124.00
#16 Cotton(80ac) Kern County	0.5	230 lbs	$126.50

* Bales of harvested hay were converted to tonnages. The bales on most of INFORM's fields weighed about 125 pounds apiece.
† Alfalfa prices were estimated conservatively at $80 per ton. During some parts of each season these prices were $90 or more per ton. Cotton prices were estimated at $0.55 per pound. On this field, the farmer conducted his own test of cotton yields, as described in Appendix A, #16.

Combined Benefits of Lower Water Use and Higher Yields

Combined benefits of $25.00 to $165.11 per acre were achieved on the six fields where both the water reductions and the yield-quantity increases were measured. (See Table 7.)

TABLE 7. COMBINED WATER-COST SAVINGS AND VALUES OF INCREASED YIELDS ON INFORM'S FIELD STRIPS

Field (See Appendix A)	Per-acre value of water-cost reductions	Per-acre yield increases	Total value of per-acre benefits
#4 Alfalfa(80ac) Yolo County-84	$5.00	$20.00	$25.00
#2 Alfalfa(38ac) Yolo County-86	$9.96	$16.00	$25.96
#28 Alfalfa(100ac) Yolo County-85	$22.50	$5.60	$28.10
#7 Alfalfa(80ac) Yolo County-86	$39.00	$16.00	$55.00
#23 Alfalfa(27ac) Fresno County-86	$8.82	$88.00	$96.82
#16 Cotton(80ac) Kern County-86	$38.61	$126.50	$165.11

Why Labor Costs and Savings Were Not Analyzed in Detail

INFORM estimated that farmers in its study spent from $1.50 to $3.00 per acre per irrigation for labor. Thus on the INFORM strips receiving one to three fewer irrigations, labor savings probably ranged from $1.50 to $9.00 per acre. However, INFORM did not attempt to analyze these savings in detail for each field strip tested because of the following complexities:

• Many hired irrigators worked in several jobs at once, including but not limited to, irrigating INFORM's test fields.

• Irrigators' skills and pay varied markedly from farm to farm, hence irrigators required more or less supervision by farm managers or owners.

Part V: Benefits Achieved

Myron Fagundes of Tulare County plans to verify INFORM's 1986 findings that about 30% less water produced over 30% more alfalfa on test strips in each of his 160-acre alfalfa fields. If practices used on INFORM's test strips are successfully extended to the entire fields, Fagundes stands to hike his per-field earnings by more than $20,000 a year.

• The use of slower and longer irrigations on some siphon-irrigated fields would not show up as labor-cost savings unless extended to entire fields. If used on this scale, however, such improvements could significantly reduce the need for costlier night labor to change siphons from one ditch location to another and/or reduce the frequency of time-consuming siphon changes.

Unmeasured Benefits of Soil-Moisture Monitoring

INFORM's data about moisture levels in crop-root zones provided cooperators with many benefits that were not tested systematically and/or could not be assigned dollar values. These benefits included:

• Higher yield quality, especially for hay and tomatoes

• Reduced damage to yields caused by overwatering some field sections on four fields

• Better information about soil types and conditions and their locations on eight fields

• Better monitoring and evaluation of soil-treatment and cultivation practices intended to increase water penetration into crop-root zones on two fields

• A reliable method for evaluating the benefits of high-tech surge irrigation valves in a pipeline on the owner's strip of one cotton field, and comparing them with benefits achieved on INFORM's strip where gypsum-block data were used to manage irrigations with standard gated pipelines.

TABLE 8. COMPARISON OF PER-SEASON COSTS OF INFORM'S TEST METHOD AND FARMER'S STREAMLINED METHOD

	INFORM's test method	Farmer's streamlined method
Number of stations on a 40-acre field	12	4
Number of blocks per station	4	2
Cost of blocks	$240.00	$40.00
Labor time for installation	6 hours	1 1/2 hours
Labor time for readings*	15 hours	5 1/2 hours
Total cost labor time @$50 for every 6 hours	$175.00	$58.33
Total costs for blocks and labor	$415.00	$98.33
Per-acre costs for blocks and labor	$10.38	$2.46†
Repairs for meter and auger and hammer set		$.50†

* Not counting time for travel to fields.
† This is the cost cited by Ron Timothy for each field monitored. But since he uses the results to plan irrigations on another, similar field, he halves the cost to $1.23 per acre for labor and blocks and 25¢ per acre for equipment repairs, for a total of $1.48 per acre.

The Declining Costs of the Gypsum-Block Method from Testing to Broad Application

The $10.38 per-acre costs of INFORM's 1985 tests on one 40-acre tomato field have dropped in three seasons to about $1.50 per acre for 600 acres of processing tomatoes, as shown in Table 8, because cooperator Ron Timothy has streamlined INFORM's method as follows:

• On each 40-acre field, Timothy monitors only TOP and END sections on each of two field strips because these locations are where uneven water distribution is likely to occur.

• At each monitored site he uses only one station be

cause experience has taught him the reliability and consistency of block readings.

- At each station he uses only two blocks to monitor 2 and 3-foot depths which give him the information he uses for scheduling irrigations.

- Data from each monitored field are collected only once a week and, in general, are used to manage irrigations on one or more unmonitored fields having similar soil types.

Timothy does not include in his cost calculations his purchase of an impedance meter at $200 and an auger and hammer set at $175. If, as he expects, these tools last for at least a decade on 600 acres of tomatoes, their costs will fall to 6¢ per acre per season. However, Timothy does include repair costs for these tools at 25¢ per acre per season.

Timothy realizes benefits of $32 per acre per season in cost savings for water and labor by using his streamlined soil-moisture method on 600 acres of tomatoes. He applies one fewer irrigation per season ("conservatively estimated") compared with his former practices, and he reduces his pumping bills by about $17 per acre. He also reduces his labor costs by $14 per acre. Half of these labor savings comes from the eliminated irrigation, and the other half comes from reducing night labor and labor for changing siphons: Timothy's slower, lower-volume irrigations allow him to use more siphons simultaneously in a field's supply ditch, and he can often water an entire 40-acre field with one "set" of siphons. Timothy's longer irrigations, running 48 to 72 hours each, reduce the need for night labor to check and change siphons.

Timothy's after-cost benefits of using the gypsum-block method on 600 acres of tomatoes amount to $30.50 per acre. His net gain is $18,300 per season.

Appendices

A
Detailed Results of INFORM's Demonstrations on 32 Fields

Sacramento Valley

Bryan Barrios

Route 2, Box 861
Woodland, CA 95695

(1) Field	Tomatoes, 35 acres
Soil type	Myers clay and Capay silty clay
County	Yolo
Year	1986
Irrigation system	Furrow
Application method	Siphons
Water source	Union School Slough. This drain ditch carrying both runoff from upslope fields and unused excess water from a federal project ran along one edge of the field.
Water cost	Less than $1 per acre-foot to pump the

Appendices

	water from the ditch to the field. There was no charge for the water itself.
Size of monitored strips	0.25 acres
Water-distribution patterns	Uniform
Changes on INFORM's strip	For the July 6 irrigation, INFORM used one 1-inch siphon per furrow, instead of the 2-inch siphons used by the owner, and doubled the length of the irrigation from 6 to 12 hours. The 1-inch siphons emitted 9 gallons per minute (gpm) per furrow compared to the 35 gpm emitted by each 2-inch siphon. The smaller streams dried out before reaching the end of INFORM's strip. For the July 23, August 7 and September 3 irrigations, INFORM used two 1-inch siphons per furrow for 6 hours (delivering 18 gpm), and one 1-inch siphon per furrow for another 6 hours, compared to the 2-inch siphons used by the owner for 6 hours. A total of four irrigations were applied to INFORM's strip from July 6 to September 3, compared to the owner-strip's five irrigations. (See Graph 8, page 86.)
Water reductions on INFORM strip	10.4 acre-inches of water per acre were applied to INFORM's strip over four irrigations, and 24 acre-inches per acre were applied to the owner's strip over five irrigations.
Yields	No visible difference according to the owner.
(2) Field	Alfalfa, 38 acres
Soil type	Yolo silty loam
County	Yolo

A: Detailed Results of INFORM's Demonstrations on 32 Fields

Year	1986
Irrigation system	Border-strip
Application method	Siphons, and cutouts in the delivery ditch
Water source	Pumped well water
Water cost	$12 per acre-foot
Size of monitored strips	3 acres
Water-distribution patterns	Uniform
Changes on INFORM's strip	The entire field was being irrigated once per cutting. The irrigator watered INFORM's strip through ditch cutouts for the June 7 and July 7 irrigations, and pump data indicated that 8.7 acre-inches per acre were applied by each watering. For the July 29 irrigation INFORM placed five 3-inch siphons in the ditch (that had been repaired) and for the August 29 irrigation four 3-inch siphons were used. On July 29, 6.8 acre-inches per acre were applied and on August 29, 4.9 acre-inches per acre were applied. INFORM's gypsum-block data showed that the amount of water needed to refill the root zone fell between 6.8 acre-inches per acre and 4.9 acre-inches per acre. This was the sole example of a "bracketing" procedure demonstrated during INFORM's tests. (See Graph 11, page 92.)
Water reductions on the entire field	From June through September, 29.1 acre-inches per acre of water were applied to INFORM's field strip, and 34 acre-inches per acre were applied to the owner's strip. However the minimum water application needed on INFORM's

Appendices

	strip was bracketed at about 6 acre-inches per acre per irrigation, so INFORM concluded that 24 acre-inches per acre would have been sufficient on its test strip.
Yields	6.1 tons per acre on INFORM's strip and 5.9 tons per acre on the owner's strip, for four cuttings.

Leroy Bertolero

Bertolero, Inc.
722 Tremont Road
Dixon, California

(3) Field	Tomatoes, 21 acres
Soil type	Yolo loam
County	Solano
Year tested	1986
Irrigation system	Furrow
Application method	Siphons
Water source	Pumped well water
Water cost	$17 per acre-foot
Size of monitored strips	One bed bounded by two standard furrows (i.e., two furrows that were not compacted by tractor tires)
Water-distribution patterns	Dry upper portions of the root zone under the tomato-bed centers in TOP and END sections
Changes on INFORM's strip	After INFORM's data for the season's first two irrigations showed that the shallower depths of the center-bed sections were being skipped, the farmer substituted 1-inch siphons for the 1-1/2-inch siphons that were being used on

A: Detailed Results of INFORM's Demonstrations on 32 Fields

both INFORM's and the owner's strips as well as on the rest of the field. The smaller siphons, however, did not noticeably improve water distribution during the season's third irrigation, whose duration was, like the first two, 12 hours. Thus, on July 12, INFORM extended the length of the fourth irrigation to 24 hours using the 1-inch siphons. This brought up the moisture levels at the 1-foot depth for the first time. On July 24, for the season's last irrigation, 3/4-inch siphons were used instead of the 1-inch siphons, the irrigation was further extended to 32 hours, and the distribution was nearly eliminated: only the 6-inch depth was skipped. (See Graph 6, page 80.)

Water reductions on INFORM's strip

For the July 24 irrigation, INFORM applied 4.6 acre-inches per acre compared to the 7 acre-inches per acre applied by the owner to the rest of the field. The potential savings over five irrigations was estimated, with the owner's agreement, as one acre-foot per acre.

Yields

No visible difference according to the owner.

Chester and Bill Roth

519 Maple Way
Woodland, CA 95695

(4) Field   Alfalfa, 80 acres
Soil type   Rincon and Marvin silty clay loam

117

Appendices

County	Yolo
Year tested	1984
Irrigation system	Border-strip
Application method	Siphons
Water source	Surface-delivered federal water
Water cost	$10 per acre-foot
Size of monitored strips	5.4 acres
Water-distribution patterns	The TOP and MID sections of the field remained partially dry after the owner's irrigations, while the END section remained very wet throughout the season.
Changes on INFORM's strip	Two of the four 4-inch siphons used by the owner on each check were removed from each of the two checks tested by INFORM after the first 6 hours of each irrigation. The remaining two 4-inch siphons were left in place for another 18 hours on INFORM's two checks. One irrigation per cutting was applied to INFORM's field strip for the June and July growth cycles compared to two per cutting applied to the owner's strip. (See Graph 7, pages 81-83.)
Water reductions on INFORM's strip	From May through August, 36 acre-inches of water per acre were applied to INFORM's field strip, and 42 acre-inches of water per acre applied to the owner's field strip.
Yields	1.34 tons per acre on INFORM's strip and 1.09 tons per acre on the owner's strip, for four cuttings.
(5) Field	Corn, 40 acres
Soil type	Capay silty clay
County	Yolo

A: Detailed Results of INFORM's Demonstrations on 32 Fields

Year tested	1985
Irrigation system	Furrow
Application method	Siphons
Water source	Pumped well water
Water cost	$12 per acre-foot
Size of monitored strips	0.43 acres
Water-distribution patterns	The field dried out continuously at all soil depths over the months of May, June and July, despite the five irrigations applied from May 12 to July 9. A large amount of surface runoff suggested that the problem was caused by applying the water too rapidly, rather than by an insufficient irrigation volume.
No test conducted	Communication difficulties rendered tests impossible. However, data from this field increased the farmers' interest in working more closely with INFORM in 1986.
(6) Field	Sugar beets, 40 acres
Soil type	Marvin-Rincon silty clay
County	Yolo
Year tested	1985
Irrigation system	Furrow
Application method	Siphons
Water source	Pumped well water
Water cost	$12 per acre-foot
Size of monitored strips	0.5 acres
Water-distribution patterns	Uniform
No test conducted	In mid-season, the crop was infested with a yellow virus blight and the field was abandoned by the owners. However, data from the field increased the farmers' interest in working with

Appendices

	INFORM in 1986.
(7) Field	Alfalfa, 80 acres
Soil type	Rincon silty clay loam
County	Yolo
Year Tested	1986
Irrigation system	Border-strip
Application method	Siphons
Water source	Surface-delivered federal water
Water cost	$10 per acre-foot
Size of monitored strips	2.9 acres
Water-distribution patterns	The field was so heavily overwatered that distribution patterns were impossible to detect in the top 4 feet of the root zone.
Changes on INFORM's strip	Two 5-inch siphons (delivering 450 gpm) were used to irrigate each of the two checks managed by INFORM, compared to the four 4-inch siphons (delivering 550 gpm) used on each of the owner's two checks. The duration of the irrigations was 12 hours per set on both strips. Five irrigations were applied to INFORM's field strip between April 29 and August 29. In the months of June and July, only one irrigation per cutting was applied to INFORM's strip compared to the two irrigations per cutting applied to the owner's strip. (See Graph 9, page 88.)
Water reductions on INFORM's strip	2.8 acre-feet of water per acre were applied to INFORM's field strip, and 6.7 acre-feet of water per acre were applied to the owner's field strip, from May through August.
Yields	For the June and July growth cycles,

A: Detailed Results of INFORM's Demonstrations on 32 Fields

when INFORM's field strip received one irrigation per cutting compared to the owner's two irrigations per cutting, 1.8 tons per acre were harvested from the INFORM strip and 1.7 tons per acre were harvested from the owner's strip. The yields for the five cuttings (May through September) were 4.9 tons per acre on INFORM's strip, and 4.7 tons per acre on the owner's strip.

Ron Timothy

B&T Farming
Box 878-D
Woodland, CA 95695

(8) Field	Corn, 30 acres
Soil type	Capay silty clay loam
County	Solano
Year	1984
Irrigation system	Furrow
Application method	Siphons
Water source	Pumped well water
Water cost	$17 per acre-foot
Size of monitored strips	0.6 acres
Water-distribution patterns	END sections on both strips were left partially dry after irrigations.
Changes on INFORM's strip	1-1/2 inch siphons were used on both the INFORM and the owner's field strips. However, each irrigation on INFORM's strip was applied for 12 hours compared to the 8 hours on the owner's strip. Three irrigations were applied to INFORM's field strip and six

Appendices

Water reductions on INFORM's strip	irrigations were applied to the owner's field strip during the season, not counting the early-season lay-by irrigation applied to both strips. 16 acre-inches of water per acre were applied to INFORM's field strip during the test period, compared to 22 acre-inches per acre applied to the owner's strip.
Yields	Gross yields were 5 tons per acre on INFORM's strip and 5.44 tons per acre on the owner's strip. The owner attributed the lower yields on INFORM's strip to an unintended 10-day delay in applying a June irrigation to INFORM's strip. This delay was caused by communication problems.
(9) Field	Tomatoes, 35 acres
Soil type	Capay silty clay on INFORM's strip; Yolo silty clay loam on the owner's strip
County	Solano
Year	1984
Irrigation system	Furrow
Application method	Siphons
Water source	Pumped well water
Water cost	$17 per acre-foot
Size of monitored strips	17 acres for INFORM; 18 acres for owner
Water-distribution patterns	Partially dry TOP section on the owner's strip after irrigations
Changes on INFORM's strip	1-inch siphons were used to apply water to each of INFORM's furrows, for each of four irrigations during the growing season after the lay-by

A: Detailed Results of INFORM's Demonstrations on 32 Fields

	owner imitated the changes made on the INFORM strip. However, on nearby fields he used 1-1/2 inch siphons and applied five or six irrigations during the season after the lay-by.
Water reductions on INFORM's and owner's strips	19.8-inches of water per acre were applied to both INFORM's and the owner's field strips. However, according to the owner, 20% less water was applied to the entire field than to nearby fields having similar soil types.
Yields	24 tons of tomatoes per acre were harvested from INFORM's field strip, and 27 tons per acre were harvested from the owner's field strip. The owner said that this difference was due to the lower water-holding capacity of the prominent soil type on the INFORM field strip, and he evaluated the yields overall as the "same" on nearby fields.
(10) Field	Corn, 50 acres
Soil type	Capay silty clay
County	Solano
Year	1985
Irrigation system	Furrow
Application method	Siphons
Water source	Pumped well water
Water cost	$17 per acre-foot
Size of monitored strips	0.62 acres
Water-distribution patterns	The field was so wet that irrigation patterns were hard to detect.
Changes on INFORM's strip	Two 1-1/4-inch siphons were used on both INFORM's and the owner's field strips. INFORM extended the length of each irrigation applied to its strip to 60

123

Appendices

	hours, compared to the 24-hour duration of each of the irrigations on the owner's strip. Five irrigations were applied to the INFORM strip for the season as a whole compared to the seven irrigations applied to the owner's strip, not counting the lay-by irrigations.
Water reductions on INFORM's strip	47 acre-inches of water per acre were applied to INFORM's field strip over five irrigations, and 50 acre-inches of water per acre were applied to the owner's strip over seven irrigations.
Yields	No visible difference according to the owner.
(11) Field	Tomatoes, 35 acres
Soil type	Yolo silty loam
County	Solano
Year	1985
Irrigation system	Furrow
Application method	Siphons
Water source	Pumped well water
Water cost	$17 per acre-foot
Size of monitored strips	0.65 acres
Water-distribution patterns	Partially dry 3-foot depth in the TOP section on the owner's strip and partially dry END section at all depths on INFORM's strip
Changes on INFORM's strip	1-inch siphons were used on both INFORM's and the owner's field strips. INFORM reduced the length of the second of the four irrigations of the season (not counting lay-by) to 30 hours, compared to the owner's 60-hour irrigation. Due to the results of INFORM's 1984 test, this cooperator had already made

A: Detailed Results of INFORM's Demonstrations on 32 Fields

Water reductions on INFORM's strip

Yields

substantial changes in his irrigation practices. He used small siphons and irrigated for up to 72 hours per set on many of his fields. INFORM's 1985 test showed that this owner could continue to refine his practices, by reducing somewhat the length of at least one of his long irrigations.

17.8 acre-inches per acre were applied to INFORM's strip, and 22 acre-inches per acre were applied to the owner's strip.

Yields on the entire field were very high, averaging 32 tons per acre. Even more important, the sugar-solids content of the crop was also very high (6.2%). Slightly less tonnage was harvested from the INFORM field strip due to the lower water-storage capacity of its soils. Yet this was insignificant given the high yields overall on a field where the owner had reduced both the volume and the number of irrigations, due to lessons learned during INFORM's 1984 tests.

San Joaquin Valley

Tony Barcello

5486 Excelsior
Hanford, CA 93230

(12) Field Cotton, 38 acres
Soil type Nord fine sandy loam

Appendices

County	Kings
Year	1986
Irrigation system	Furrow/border-strip. Using this hybrid system, the owner divided the field into six sections of 6.3 acres each. Berms (low levees) divided the sections, and ditches for carrying water from the top of the field to the drain ditch were dug on each side of the berms. On each 6.3-acre section, furrows were cut into a level plane running perpendicular to the sloped berms and ditches as illustrated on the facing page.
	This system requires big streams of water to fill the furrows and thus results in very rapid irrigations. Its advantages are that it saves on the cost of siphons (or gated pipeline) and on labor. Its disadvantages are that it takes more work to level and prepare the field; it provides less control over water applications; and, due to the rapid, heavy applications, water often piles up at the end of the field.
Application method	Hydrants (one per check)
Water source	People's Ditch Company
Water cost	Less than $5 per acre per season
Size of monitored strips	6.3 acres
Water-distribution patterns	Partially dry TOP sections on both strips at all depths
Test invalidated	Records kept by the owner showed significant differences in the amounts of water applied to each of the six field sections for each irrigation. Five irrigations were applied to INFORM's strip during the growing season, and six irri-

A: Detailed Results of INFORM's Demonstrations on 32 Fields

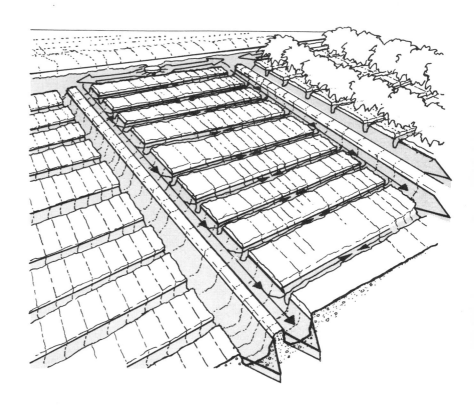

gations were applied on the owner's strip. Nonetheless, the total volume of water applied to the INFORM strip was 18.75 acre-inches per acre, compared to the 14.1 acre-inches per acre applied to the owner's strip.

Myron Fagundes

7323 Elder Avenue
Hanford, CA 93230

(13) Field   Alfalfa, 160 acres (north)
Soil type   Remnoy-Youd fine sandy loam

Appendices

County	Tulare
Year	1985, 1986
Irrigation system	Border-strip
Application method	Hydrants (one per check)
Water source	Pumped well water (diesel)
Water cost	$20 per acre-foot
Size of monitored strips	12.4 acres
Water distribution patterns	In 1985, the field was so heavily overwatered that no distribution patterns were visible in the top 4 feet of the root zone monitored by gypsum-block stations. In 1986, the owner reduced the volume of each irrigation on both INFORM's and the owner's strips (independently of INFORM's tests). On both field strips, these lighter applications left some portions of the root zone on the END field sections partially dry. (See Appendix B.)
Changes on INFORM's strip in 1986 test	Five irrigations were applied to INFORM's field strip, and seven irrigations were applied to the owner's field strip, between May 1 and August 25. In June and July, one irrigation per cutting was applied to INFORM's strip compared to the two irrigations per cutting applied to the owner's strip.
Water reductions on INFORM's strip	No estimates could be made for the hydrants' water-outflow rates. However, according to the owner, all the irrigations of the season were roughly equal in volume and duration. Therefore, INFORM estimated that 29% less water was applied to INFORM's strip

A: Detailed Results of INFORM's Demonstrations on 32 Fields

Yields

than to the owner's strip during the test period.

In 1986, yields for four cuttings were compared. For the two cuttings after both INFORM's and the owner's strips had been equally watered (May 23 and August 25), yields were 5% higher on INFORM's strip. For the two cuttings after INFORM's strip had been watered once and the owner's strip had been watered twice (June 23 and July 24) yields were 30% higher on INFORM's strip. For these two cuttings, 3.69 tons per acre were harvested from INFORM's strip and 2.64 tons per acre were harvested from the owner's strip.

In June and July 1985, two irrigations per cutting on INFORM's strip produced 2.57 tons per acre for the two cuttings combined. In June and July 1986, when the same field strip was watered only once per cutting, it produced 3.69 tons per acre for these two cuttings.

(14) Field	Alfalfa, 160 Acres (south)
Soil type	Remnoy-Youd fine sandy loam
County	Tulare
Year	1985, 1986
Irrigation system	Border-strip
Application method	Hydrants (one per check)
Water source	Pumped well water
Water cost	$20 per acre-foot
Size of monitored strips	12.4 acres
Water-distribution patterns	In 1985, the field was so heavily overwatered that no distribution patterns

Changes on INFORM's and owner's strips in 1986

were visible in the top four feet of the root zone monitored by gypsum-block stations. In 1986, the owner reduced the volume of each irrigation on both INFORM's and the owner's strip (independently of INFORM's tests), and a uniform pattern of water distribution was revealed.

Five irrigations were applied to both strips between May 2 and August 20. This amounted to two fewer irrigations than were applied to the rest of the field. For the May 23 to June 19 growth cycle, the owner applied only one irrigation to both strips. For the June 19 to July 21 cycle, the owner applied two irrigations to INFORM's strip and only one irrigation to the owner's strip (due to some confusion about the directions). For the July 21 to August 20 cycle he applied one irrigation to INFORM's strip and two irrigations to the owner's strip. Thus, both INFORM's and the owner's strips were tested on a one-irrigation regime for one growth cycle each during the hottest part of the season.

Water reductions on INFORM's strip

No estimates could be made for the hydrants' water-outflow rates. However, based on the farmer's statement that his applications were about equal in volume and length, INFORM estimated that the applications on both INFORM's and the owner's field strips were 29% less than those on the re-

A: Detailed Results of INFORM's Demonstrations on 32 Fields

Yields

mainder of the field for the 4-month period tested.

Yield results for 1986 are difficult to interpret because data from the June cutting are missing. For three cuttings (May 23, July 21, and August 20) 5.75 tons per acre were harvested on the INFORM strip (which was watered four times during the three growth cycles) and 4.20 tons per acre were harvested on the owner's strip (which was also watered four times during the three cycles).

A comparison of 1985 and 1986 yield data from the INFORM field strip, however, suggests marked yield increases due to the fewer and lighter irrigations applied to this strip in 1986. In 1985, the combined yields of the May, July and August cuttings on this strip were 4.19 tons per acre with a total of six irrigations. By contrast, in 1986 the three comparable cuttings yielded 5.75 tons per acre with a total of four irrigations.

Craig Fulwyler

McConnell Farms
130 7th Street
Wasco, CA 93280

(15) Field   Alfalfa, 80 acres
Soil type   Milham fine sandy loam
County   Kern

Appendices

Year	1985, 1986
Irrigation system	Border-strip
Application method	Hydrants (one per check)
Water source	Semi-Tropic Water Storage District
Water cost	$33 per acre-foot (including the water-purchase price and the cost of pumping the water from the large delivery canal to the field)
Size of monitored strips	2.9 acres
Water-distribution patterns	Uniform (See Graph 3, page 72.)
No test conducted	This owner was already irrigating his alfalfa field only once per cutting and was achieving high annual yields of 10 tons per acre in 1985 and 1986 for seven cuttings.
(16) Field	Cotton, 80 acres
Soil type	Milham fine sandy loam
County	Kern
Year	1986
Irrigation system	Furrow
Application method	Gated pipeline
Water source	Semi-tropic Irrigation District
Water cost	$33 per acre-foot (including the water purchase price and the cost of pumping it to the field)
Size of monitored strips	0.5 acres
Water-distribution patterns	Partially dry MID and END sections on both strips after irrigations
Changes on INFORM's and owner's strips	The first irrigation of the regular season (applied May 25 to the rest of the field) was deliberately skipped on INFORM's strip and mistakenly skipped on the owner's strip. The fifth irrigation of the season (applied the third week of

A: Detailed Results of INFORM's Demonstrations on 32 Fields

August to the rest of the field) was skipped on the INFORM field strip only.

Altogether four irrigations were applied to INFORM's field strip, and five irrigations were applied to the owner's field strip, compared to the six irrigations that were applied to the rest of the field.

Water reductions on INFORM's strip

From the first of May to the first of September, 35 acre-inches of water per acre were applied to INFORM's field strip, and 42 acre-inches of water per acre were applied to the owner's strip; 49 acre-inches of water per acre were applied to the rest of the field.

Yields

The owner developed and carried out his own test of the yield differences on INFORM's and the owner's field strips, as illustrated in the drawing on the following page. First, he harvested and weighed each of the two four-row strips separately (A). Then he harvested and added together the yields from the two rows bordering each tested strip (B). Finally, he harvested and added together the yields from the two rows on each side of the bordering strips, which were far enough away to be unaffected by a "border effect" (C). Thus he developed three comparisons per strip.

The results of these yield tests were as follows:

Appendices

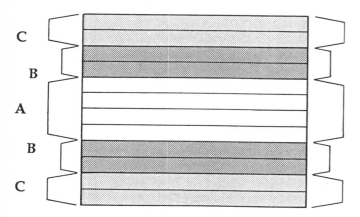

A = Four tested beds
B = Four bordering beds
C = Four beds adjacent to bordering beds

INFORM strip
(four irrigations)
A. 2.56 bales/acre—tested rows
B. 2.29 bales/acre—border rows
C. 2.15 bales/acre—normal nearby field rows

Owner strip
(five irrigations)
A. 2.67 bales/acre—tested rows
B. 2.65 bales/acre—border rows
C. 2.79 bales/acre—normal nearby field rows

The yields on the INFORM test strip were thus discovered to be *.41 bales per acre higher than on the nearby field sections* which received two more irrigations during the season.

However, the yields on the owner's strip were discovered to be *0.12 bales per acre lower than the nearby field sections* which had received one fewer irrigation during the season.

This ingenious test shows some of the

A: Detailed Results of INFORM's Demonstrations on 32 Fields

complexities involved in comparing yields between widely separated field strips.

The INFORM strip was located in a field section which produced on average 0.64 bales per acre less than the section in which the owner's strip was located. The owner's evidence, however, indicated that reductions in water applications on the INFORM strip helped to overcome some of the impact of other field conditions, especially less fertile soil, that accounted for lower yields here than on other field sections.

Randy Gafner

Oran Gil Farms, Inc.
1420 7th Street
Wasco, CA 93280

(17) Field	Alfalfa, 80 acres
Soil type	Nahrub clay lethent complex IIIs-6 (salinity >8, soil pH >8.4)
County	Kern
Year	1985, 1986
Irrigation system	Border-strip
Application method	Hydrants
Water source	Pumped well water (pH=10)
Water cost	$43.82 per acre-foot
Size of monitored strips	1.9 acres
Water-distribution patterns	Severe water penetration problems on TOP and MID sections on both INFORM's and the owner's field strips due to saline and alkaline conditions

Appendices

Changes on INFORM's and owner's strips

In 1986, sulfurized water was used to irrigate both INFORM's and the owner's field strips in an effort to reduce water-penetration problems caused by the complex chemistry of clay soils, salt buildups and alkaline irrigation water. After six sulfur treatments, irrigation water was shown by gypsum-block readings to penetrate to the two-foot depth on field sections where, in 1985, it had not even reached the 1-foot depth. (See Graph 4 and text, pages 73-76.)

In July 1986, one irrigation was applied during the growth cycle on the INFORM strip, and two irrigations were applied to the owner's strip. A total of seven irrigations were applied to the INFORM strip from May through September and eight irrigations were applied on the owner's strip.

Water reductions on INFORM's strip

Thus, from May through September, about 49 acre-inches of water per acre were applied to INFORM's strip, and 56 acre-inches per acre were applied to the owner's strip.

Yields

During the July growth cycle, when INFORM's strip received only one irrigation, 12 bales of alfalfa were harvested from this strip, compared to the 24 bales harvested from the owner's twice-watered strip. These lower yields were due primarily to the greatly restricted water- storage capacity of the blighted TOP and MID sections of the field on the INFORM strip. (The same

A: Detailed Results of INFORM's Demonstrations on 32 Fields

conditions on the owner's strip were partially offset by using two irrigations.) For the four cuttings from April through July, INFORM's strip produced 3.4 tons per acre, and the owner's strip produced 4 tons per acre.

(18) Field	Cotton, 120 acres
Soil type	Nahrub clay with alkali-spots
County	Kern
Year	1986
Irrigation system	Furrow
Application method	Gated pipeline
Water source	Pumped well water
Water cost	$40 per acre-foot
Size of monitored strips	0.5 acres
Water-distribution patterns	Partially dry TOP and MID sections on both strips after irrigations in July and August
Test Invalidated	Four irrigations were applied to INFORM's field strip, and six irrigations were applied to the owner's strip between the end of May and the end of August.

However, the results of these tests were discounted when INFORM learned that the farmer had applied a pre-irrigation to the field section that included the owner's strip but not to the section that included INFORM's strip. Thus the young crop on the owner's strip received critically important additional moisture early in the growing season, and yields were better on the owner's strip.

Appendices

Bill Longfellow
823 Yosemite Drive
Hanford, CA 93230

(19) Field	Alfalfa, 40 acres
Soil type	Kimberlina fine sandy loam IIs-6
County	Kings
Year	1985, 1986
Irrigation system	Border-strip
Application method	Hydrants (one per check)
Water source	Pumped well water (plus dairy waste)
Water cost	$27.00 per acre-foot
Size of monitored strips	7 acres
Water-distribution patterns	The END section on INFORM's strip had a severe problem of water penetration which was eventually discovered to be due to the layering of different soil types during a recent releveling of the field. (See pages 76-77.)
	In 1985 and 1986 the owner's field strip was so heavily watered that distribution patterns were impossible to detect in the top 4 feet of the root zone that were monitored by gypsum blocks.
Changes on INFORM's and owner's strips	In 1986, on INFORM's strip, a larger volume of water was applied during one July 10 irrigation. However, this additional application did not reduce the problem of water penetration occurring in the END section of INFORM's field strip.
	In both June and August, only one irrigation per cutting was applied to both INFORM's and the owner's field strips, compared to the two irrigations per cutting applied to the rest of the

field. (There was some confusion in giving or receiving instructions: Directions intended only for the INFORM field strip were applied to both INFORM's and the owner's strips.)

Water reductions on INFORM's and owner's strips

Only six irrigations were applied to both INFORM's and the owner's field strips from May through August, compared to the eight irrigations applied to the remaining field sections.

No estimates could be made for the hydrants' water-outflow rates. Also, because each irrigation on each check was managed somewhat differently by the owner, INFORM could not assume that each irrigation was roughly equal in volume. Thus no estimates were made of the percentage by which water use was reduced on INFORM's and the owner's strips.

Yields

Despite the problem of water penetration on INFORM's strip, 1.47 tons per acre were harvested in June after the strip received one irrigation, but only 1.07 tons per acre were harvested in July after two irrigations to the strip. Moreover, despite problems of water penetration, INFORM's yields were higher than those on the owner's strip when each strip was given one irrigation per cutting cycle, in the months of June and August. The yields were 1.47 tons and 0.96 tons per acre on INFORM's strip, and 0.82 and 0.88 tons per acre on the owner's strip.

Finally, on the owner's strip, bale

Appendices

counts were lower when the field was watered once per cutting than when it was watered twice. For example, the yield for the June growing cycle was 0.82 tons per acre (with one irrigation) compared to 0.97 tons per acre in July (with two irrigations).

These results on the owner's strip do not appear to have been influenced by soil-moisture levels because these were maintained at field capacity for the entire growing season, with the exception of occasional drops at the 1-foot depth.

The farmer sent hay samples from the drier INFORM strip and from one of the wetter field sections for laboratory testing of nutrient content. The hay grown on the dried strip was found to have "substantially" more nutritive value than hay from the wetter sfield section.

(20) Field	Cotton, 40 acres
Soil type	Kimberlina fine sandy loam
County	Kings
Year	1985
Irrigation system	Combination furrow and border-strip (see Field 12)
Application method	Hydrants
Water source	Pumped well water
Water cost	$27 per acre-foot
Size of monitored strips	8 acres
Water-distribution patterns	The field remained continuously saturated at all four depths monitored from the first of July and until the mid-

A: Detailed Results of INFORM's Demonstrations on 32 Fields

No test conducted	August cutoff. The entire season was devoted to field observations and discussions with owner.
(21) Field	Cotton, 40 acres
Soil type	Kimberlina fine sandy loam with a sandy streak running diagonally through the field where an old slough had been filled in
County	Kings
Year	1986
Irrigation system	Furrow/border-strip (see Field 12)
Application method	Hydrants (one per check)
Water source	Pumped well water
Water cost	$27
Size of monitored strips	7 acres
Water-distribution patterns	The TOP and MID sections of INFORM's strip were left partially dry by irrigations in July and August. (Note: the farmer's combination border-strip/furrow irrigation system precluded tests of slower application rates that might have overcome this problem.)
Changes on INFORM's strip	The spacing of irrigations was lengthened between July 20 and the end of August, so that three irrigations were applied to the INFORM strip compared to the four irrigations applied to the owner's strip. For the season as a whole, four irrigations were applied to INFORM's strip and five irrigations were applied to the owner's strip.

Appendices

Water reductions on INFORM's strip	No estimates were made for the percentage by which water use had been reduced on INFORM's strip, for the same reason given for Field 19, above, which was owned by the same farmer.
Yields	No observable difference.
(22) Field	Cotton, 160 acres
Soil type	Kimberlina fine sandy loam
County	Kings
Year	1986
Irrigation system	Furrow/border strip (see Field 12)
Application method	Hydrants (one per check)
Water source	Lakeside Irrigation District
Water cost	$17 per acre-foot
Size of monitored strips	14 acres
Water-distribution pattern	Frequent, heavy watering obscured distribution patterns.
No test conducted	The owner preferred not to test this field for water reductions because he considered high-frequency irrigation necessary for high yields.

Delbert Mello

2258 East Riverdale Avenue
Laton, CA 93242

(23) Field	Alfalfa, 27 acres
Soil type	Traver fine sandy loam (alkaline conditions: pH=8 -10)
County	Fresno
Year	1986
Irrigation system	Border-strip
Application method	Hydrants (one per check)

A: Detailed Results of INFORM's Demonstrations on 32 Fields

Water source	Pumped well water
Water cost	$14 per acre foot
Size of monitored strips	1.7 acres
Water-distribution patterns	The END section on INFORM's strip was left partially dry after irrigations due to saline and alkaline conditions. The owner's field strip was so heavily watered that no distribution patterns were possible to detect for the first two months of the growing season. After that, only the top foot of the root zone dried out at any point down the run, and this change was too slight to indicate distribution patterns.
Changes on INFORM's strip	During the July and August growth cycles, only one irrigation per cutting was applied to INFORM's strip compared to the two irrigations per cutting applied to the owner's strip. For the May and June cycles, two irrigations were applied to both INFORM's and the owner's field strips.
Water reductions on INFORM's strip	From May through August, 26.6 acre-inches of water per acre were applied to the INFORM's strip, and 34.2 acre-inches per acre were applied to the owner's strip.
Yields	For every cutting of the test period, May through August, the yields on INFORM's strip were higher than on the owner's strip, despite the problems of water penetration on the INFORM strip. 8.6 tons per acre were harvested from INFORM's strip and 7.5 tons per

Appendices

acre were harvested from the owner's strip, for four cuttings.

Gary Wilson

Wilson Ag
P.O. Box 728
Shafter, CA 93263

(24) Field	Skip-row cotton, 80 acres (four rows planted, four rows fallow)
Soil type	Panoche clay loam, Garces silty loam, IIIs-6
County	Kern
Year	1985, 1986
Irrigation system	Furrow
Application method	Gated pipeline on INFORM's strip; surge-valve pipeline on the owner's strip
Water source	Pumped well water
Water cost	$40 per acre-foot
Size of monitored strips	0.4 acres
Water-distribution patterns	Uniform. On this field in 1986, surge-irrigation equipment was used on the owner's strip, while regular gated-pipeline equipment was used on the INFORM strip. The owner had spent about $65 per acre for this equipment and was testing its use. Gypsum-block data revealed no significant difference between the water distribution achieved with the new equipment from that achieved with the old.
Changes on INFORM's strip	*1985 Test*: The second and sixth irrigations of the season were skipped

A: Detailed Results of INFORM's Demonstrations on 32 Fields

on INFORM's strip. In all, four irrigations were applied to INFORM's strip and six irrigations were used on the owner strip.

1986 Test: The first, sixth, and seventh irrigations of the season were skipped on INFORM's strip. Altogether four irrigations were applied to INFORM's strip, and seven irrigations were applied to the owner's strip.

Pre-irrigations: INFORM also monitored (1985) and tested (1986) pre-irrigations on this field. The 1985 gypsum-block readings indicated that an application of 3 acre-inches per acre would have restored the root zone to capacity before planting. The owner applied 14 acre-inches per acre to the field.

In 1986, the gypsum-block readings indicated that 5 acre-inches per acre were needed for full root-zone replenishment, and this amount was applied to INFORM's strip. The farmer applied 11 acre-inches per acre to his strip.

Water reductions on INFORM's strip

1985: 36-acre-inches of water per acre were applied to INFORM's strip from May through August, and 50 acre-inches per acre were applied to the owner's field strip, not counting the pre-irrigation.

1986: 37 acre-inches of water per acre were applied to INFORM's strip, *including* the 5 acre-inches applied during pre-irrigation, and 45 acre-inches per acre were applied to the owner's strip, *including* the 10.5 acre-inches applied during the pre-irrigation.

Not counting the pre-irrigations, 32 acre-inches of water per acre were applied to INFORM's strip and 34.5 acre-inches per acre were applied to the owner's strip. The owner's water use had dropped markedly from 1985 to 1986 due to his purchase of a new pipeline having surge valves that forced water through the gates in spurts. INFORM achieved similar results using a much cheaper method.

Yields

1985 Test: According to the farmer, the yields on INFORM's strip in 1985 were enough lower that the cost savings achieved by INFORM's water reductions did not offset the revenue loss. However, the farmer admitted that an extended delay in applying a June irrigation, and not an insufficient irrigation volume, was probably responsible for this yield decrease. The delay was caused by logistical problems of irrigating INFORM's strip separately from the rest of the field.

1986 Test: The farmer saw no difference between the yields on the INFORM strip and the yields on the owner's strip.

A: Detailed Results of INFORM's Demonstrations on 32 Fields

Wayne Wisecarver
Wisecarver Farms
15519 Fifth Avenue
Hanford, CA 93230

(25) Field	Alfalfa, 80 acres
Soil type	Garces loam and Grangeville fine sandy loam
County	Kings
Year	1985, 1986
Irrigation system	Border-strip
Application method	Hydrants
Water source	Pumped well water
Water cost	$23 per acre-foot
Size of monitored Strips	12.3 acres (This 80-acre field was irrigated the "long way" and the run was half a mile in length.)
Water-distribution patterns	The END of the field on the INFORM strip was insufficiently irrigated, and remained partially dry throughout the latter part of the season. The MID section on the owner's field strip had a lower water-holding capacity than the rest of the field (due to alkali "hot" spots) and remained partially dry after irrigations throughout the season.
Changes on INFORM's and owner's strips	In 1986, both the owner's and INFORM's field strips were tested, during one growth cycle each, by reducing the number of irrigations from two to one. Thus between May 6 and August 20, six irrigations were applied to each strip compared to the seven irrigations applied to the rest of the field.
Water reductions	No estimates could be made for the hydrants' water-outflow rates. How-

147

Appendices

Yields

ever, based on the farmer's statement that his applications were about equal in length and volume throughout the season, INFORM estimated that 14% less water was applied to both INFORM's and the owner's strips than to the rest of the field.

Yields were lower on both field strips when they were watered once instead of twice. However, in both cases, these results appeared to be influenced by the problems of uneven water distribution.

(26) Field — Cotton, 40 acres
Soil type — Kimberlina fine sandy loam
County — Kings
Year — 1985
Irrigation system — Furrow/border-strip (See Field 12.)
Application method — Hydrants (one per check)
Water source — Pumped well water
Water cost — $23 per acre-foot
Size of monitored field — 7 acres
Water-distribution pattern — Uniform
No test conducted

As was true with all but one other field analyzed in 1985 in the San Joaquin Valley, the full season was used to collect data and discuss them with owners, before asking for their cooperation in testing irrigation changes the next season.

(27) Field — Cotton, 52 acres
Soil type — Kimberlina fine sandy loam, Nord fine sandy loam
County — Kings
Year — 1986

A: Detailed Results of INFORM's Demonstrations on 32 Fields

Irrigation system	Furrow/border-strip (See Field 12.)
Application method	Hydrants (one per check)
Water source	Lakeside Irrigation District (federal water)
Water cost	$17 acre-foot
Size of monitored strips	0.5 acres
Water-distribution pattern	The TOP section on INFORM's strip dried out during the season, while the MID and END sections were continuously saturated at all four depths tested. The 2-foot soil depth on the owner's field strip was skipped by every irrigation of the season, except the last.
Changes on INFORM's strip	Four irrigations were applied to INFORM's strip after the lay-by, compared to five irrigations applied to the owner's field strip.
Water reductions on INFORM's strip	No estimates could be made for the hydrants' water-outflow rates. However, based on the farmer's statement that his applications were about equal in length and volume throughout the season, INFORM estimated that 20% less water was applied to both INFORM's and the owner's strips than to the rest of the field.
Yields (not comparable)	The owner reported no visible difference in yields, but added that the two sides of the field generally produce very different yields due to different soil types.

Appendices

Four Fields Belonging to Cooperators Who Withdrew from INFORM's Study After One Season

(28) Field	Alfalfa, 100 acres
Soil type	Yolo silty loam
County	Yolo
Year	1985
Irrigation system	Border-strip
Application method	Siphons and cutouts in delivery ditch
Water source	Pumped well water
Water cost	$22.50 acre-foot
Size of monitored strips	3.7 acres
Water-distribution patterns	Partially dry END section on owner's strip after irrigations
Changes on INFORM's strip	INFORM used one irrigation compared to the owner's two irrigations for the August growing cycle. Both the INFORM and the grower used only one irrigation in September.
Water reductions on INFORM's strip	In August, the single INFORM irrigation and the first of the owner's two irrigations were "wild" and could not be measured. (Sections of the delivery ditch were cut out and water was allowed to spill freely onto the field.) For the second August irrigation INFORM persuaded the farmer to use siphons on the owner's strip and he applied 9 acre-inches of water per acre. Assuming that the wild irrigations on both strips were similar in volume, INFORM's strip received 9 acre-inches per acre less water during the August cutting cycle than the owner's strip. In September, siphons were used on both strips, and only one full irrigation

A: Detailed Results of INFORM's Demonstrations on 32 Fields

was applied to each: 6 acre-inches per acre were applied to INFORM's strip, and 9 acre-inches per acre were applied to the owner's strip.

The total reductions on the INFORM strip amounted to one acre-foot per acre over two irrigations.

Yields — 1.69 tons per acre on INFORM's strip in August when it was watered once, and 1.62 tons per acre on the owner's strip which was watered twice.

(29) Field	Tomatoes, 120 acres
Soil type	Reiff fine sandy loam
County	Yolo
Year	1985
Irrigation system	Furrow (sideroll sprinklers were used for germinating the crop in April)
Application method	Siphons
Water source	Pumped well water
Size of monitored strips	1.6 acres
Water-distribution patterns	This field dried out continuously throughout the growing season, despite the eight irrigations applied. The 1-1/4 inch siphons used were too small to send streams down the 2,500-foot-long furrows during the 12-hour sets. Only when these siphons were left in place for 24 hours on INFORM's strip (on June 24 and July 14) did the END of the field show any increase in moisture levels.
No test conducted	It was impossible to make contact with the owner to arrange for the tests.

151

Appendices

(30) Field	Alfalfa, 40 acres
Soil type	Kimberlina fine sandy loam
County	Kings
Year	1985
Irrigation system	Border-strip
Application method	Hydrants (one per check)
Water source	Pumped well water
Water cost	Not available
Water distribution patterns	Partially dry TOP and MID sections on the owner's field strip and very dry END sections on INFORM's field strip that may have been caused by an impermeable soil layer created during the field's recent releveling
No test conducted	The cooperator chose not to participate in the study in 1986, when a water-reduction test was planned.
(31) Field	Cotton, 60 acres
Soil type	Kimberlina fine sandy loam, Nord fine sandy loam
County	Kings
Year	1985
Irrigation System	Furrow
Application method	Gated pipeline
Water source	Pumped well water
Water cost	Not available
Size of monitored strips	0.75 acres
Water distribution patterns	Frequent irrigation obscured distribution patterns.
No test conducted	The cooperator chose not to participate in the study in 1986, when a water-reduction test was planned.

A: Detailed Results of INFORM's Demonstrations on 32 Fields

One Sprinkler Irrigated Field

Harry Dewey

P.O. Box 203
Yolo, CA 95697

(32) Field	Almond orchard, 20 acres
Soil type	Yolo silty loam
County	Solano
Year	1984
Irrigation system and application method	Solid-set sprinkler
Water source	Pumped well water
Water cost	$14 per acre-foot
Water-distribution patterns	No water penetration below the one-foot depth
No test conducted	Because of the sprinkler system's design, it could not be operated so as to irrigate a test strip separately from the rest of the orchard.

B
Two Additional Examples of Uneven Distribution

Two examples of water distribution showing interesting patterns come from an underirrigated tomato-field strip and an alfalfa-field strip that was overirrigated during one season and, due to lighter irrigations, underirrigated in some sections during the following season.

B: Two Additional Examples of Uneven Distribution

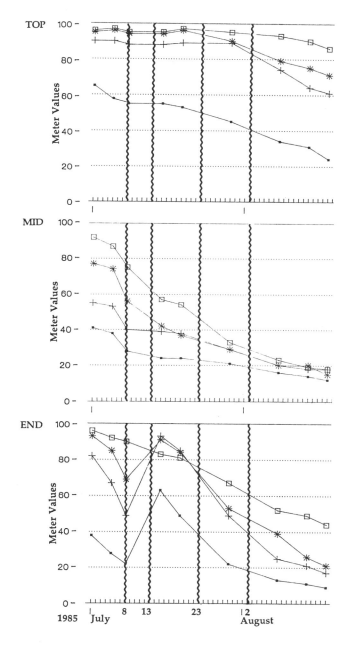

Graph 12. An Underirrigated Tomato-Field Strip

In 1985, on a 120-acre tomato field in Yolo County, the MID and END sections of the owner's strip (shown here) and INFORM's strip (not shown) dried out nearly continuously at all depths from the first of July to the August 2 cut off date, despite the four irrigations applied during this period. The reason for this drying out, which caused visible crop damage, was the use of 1-inch siphons which were too small to emit streams sufficiently large to reach the end of the 2,000-foot run.

Four monitoring sites were used on each strip in this field because of its long run. However, here, readings from the two MID sections have been averaged and displayed along with data from TOP and MID sections.

Key		1 ft	-■-
irrigation	}	2 ft	+
		3 ft	✷
		4 ft	⊟
To read meter values, see page 38.			

155

Appendices

Graph 13. Overwatering Obscures Patterns of Water Distribution on an Alfalfa-Field Strip

In 1985, INFORM's strip on this 160-acre field was kept continuously wet. Soil-moisture patterns for the entire season were similar to those shown here, where gypsum-block readings remained at 95 from July 10 to the end of the month.

In 1986, the same field strip was both tested for fewer irrigations by INFORM and also watered more lightly per irrigation by the farmer, for reasons unknown to INFORM. The farmer's lighter water applications revealed patterns of uneven distribution in the root zone on the strip's END section: The irrigation of July 6 failed to replenish completely the 3 and 4-foot soil depths.

Key	
	1 ft ■
irrigation ⟩	2 ft +
	3 ft ✶
cutting ⎮	4 ft ⊟
To read meter values, see page 38.	

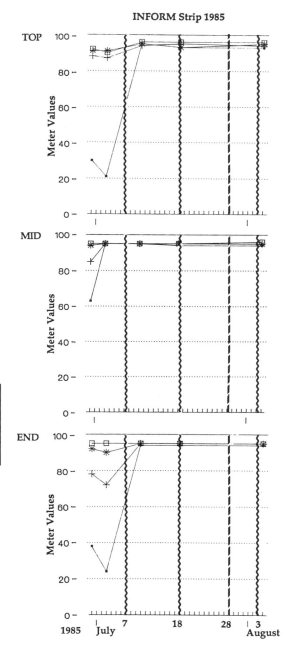

B: Two Additional Examples of Uneven Distribution

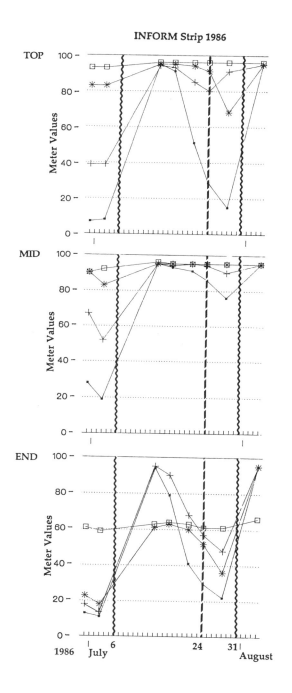

C
Why 10 Fields Were Not Tested

INFORM analyzed distribution patterns on all 32 fields in its study, but tested irrigation changes on strips of only 22 of these fields. The reasons why ten fields were not tested, given below, reflect the highly site-specific nature of INFORM's tests and the different interests and perspectives of the cooperating farmers.

• On four fields of row crops, one season's data on water distribution in the crop-root zone were used to persuade farmers to test water reductions on other fields during a second season. The row crops being analyzed were not replanted on the same fields each year.

• On two fields, the farmers, who were cooperating in INFORM's tests on other fields, chose not to conduct tests. One of these fields was already being efficiently irrigated and was producing very high yields. The other was being frequently irrigated as part of the farmer's plan to increase yields.

C: Why 10 Fields Were Not Tested

- Three fields belonged to farmers who, for different reasons, decided not to participate during a second season when water-reduction tests were planned.

- On one sprinkler-irrigated field in the study, the system's design precluded the irrigation of one test strip separately from the rest of the field.

D
Flow Rates Through Siphons

The amount of water flowing through a siphon from a delivery ditch onto a field is determined by the siphon's diameter, and by the drop in water level from the delivery ditch to the field surface (the "head"). As the following graphs show, a 3/4-inch siphon with a 6-inch head delivers 6 gallons per minute (gpm), whereas a 1-inch siphon with the same head delivers 9 gpm, and a 2-inch siphon delivers 35 gpm.

There is no tool commercially available for farmers to measure siphon heads. For INFORM's tests, however, Peter Mueller-Beilschmidt developed a simple device for the job by suspending two small floats from the ends of a short rod.

D: Flow Rates Through Siphons

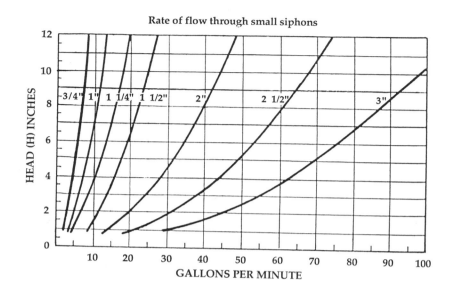

Rate of flow through small siphons

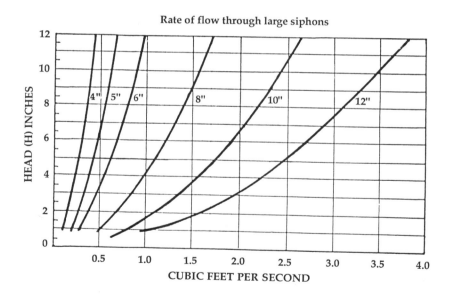

Rate of flow through large siphons

Source: Measuring Irrigation Water, Leaflet 2956, Division of Agricultural Sciences, University of California

E
How INFORM's Cooperators Evaluated The Soil-Moisture Method

In November 1986 INFORM's cooperators discussed INFORM's test results and their views of the soil-moisture method. The following quotations come from these interviews.

Randy Gafner
Kern County

When I was little, I used to go around with my Dad and I would ask him how he knew when to irrigate. And, say, at the beginning of the cotton season when the plants were just a few inches high, he'd tell me, "Those plants look thirsty, but underneath they're growing big, strong roots." And I'd wish I could look under the soil to see what was happening.

I guess what I'd say about gypsum blocks is that they're like having eyes under the soil.

E: How INFORM's Cooperators Evaluated The Soil-Moisture Method

Leroy Bertolero
Solano County

The tomato plants tell you everything. You can go out there and see that these plants are overwatered and they're wilted or these others are underwatered and are turning dark. So the tomato will tell you what you are doing right or wrong. But it's too late to get the water out of the ground, or maybe it's very difficult to get the ground rewetted without driving the oxygen out and hurting the plants. So the problem with reading the vine is that unless you are just extremely fortunate, you're going to be doing things after the fact.

The soil-moisture method allows you to know what is happening so you don't have as many problems. You know if you are building up too high a water level. Then you either have to shorten the irrigation length or delay the timing.

Craig Fulwyler
Kern County

I think I would like to pursue blocks some more. The gypsum blocks give you data and I need to associate that with the crop indicator. And I haven't got to the point where I can see how crop stress relates to soil moisture and just manage the crop that way. And I feel that gypsum blocks are a good program, a good tool, to teach me that.

Gary Wilson
Kern County

Gypsum blocks have enabled us to compare surge a little bit better. They have given us a look at three different

Appendices

Bill Roth
Yolo County

Bryan Barrios
Yolo County

Myron Fagundes
Tulare County

Ron Timothy
Solano County

points in a field which we were not doing before. They've sort of given us a side-by-side with the neutron probe. The gypsum block definitely has some merits. It's cheaper than probes and you don't have to have licenses and so forth.

I got the principle down. Don't just try to put the water on in an hour or two and then turn it off because that's no good. It won't soak down. If you push the water through with a big head, why maybe it won't soak down. You've got to let it run so many hours.

How much does all this cost? (INFORM: a meter costs $200 and the blocks cost $5.) One irrigation would just about pay for that.

Oh, we're going to pursue it some more next year. We're going to try one irrigation and see, and then we can compare again just like we did this year and do the bale counts again and see how it's going to work out.

What I'll probably do is irrigate the south field once and then the next time irrigate it twice, and the next time irrigate once, and see what the bale counts are. If we are coming pretty close production-wise and if the grass is less weedy, we may start changing over.

There's a lot about digging the holes for installing gypsum blocks. I have the guy that digs them bring me the core samples. I have him take three bags and bring back a sample from each level. That gives me a lot of information about my soil types on my different fields.

Then too, when you have a gypsum-block station in a field, it makes you go back to that spot, which is real important. If you go to the same spot all the time it gives you a better feel about what that field is doing. If you go to different areas of the field, different areas of the field do different things because the soil type changes so much in this country. You tend just to look at the good spots.

Bill Longfellow
Kings County

Well, I ran my own little experiment on 10 acres of cotton right here by the commodity barn. It went 3.1 bales this year. And I know I cut out one, possibly two, irrigations. I took the readings and scheduled out the irrigations farther. The growth was a lot shorter, but I think that's an exceptional yield for this year.

F
Letter from the Westlands Water District Evaluating Gypsum Blocks After Comparing Them with a Neutron Probe in 1986 Field Tests

March, 1988

Dear INFORM:

In 1986, the water conservation and management staff of the Westlands Water District conducted tests on three cotton fields to compare the neutron probe that we use regularly to monitor soil moisture, with the gypsum blocks used in INFORM's tests. At each of four sites on each field, we installed a neutron probe access tube about 6 inches away from an array of gypsum blocks buried at one-foot intervals to four feet. Every week for a 6-month period we collected readings from all 12 sites. These showed several things about the gypsum blocks quite clearly:

1) There was no calibration used with the gypsum blocks. The consistency between blocks appeared quite good. This is in contrast with the generally held opinion that individual calibration is required with the blocks.

F: Letter from the Westlands Water District

2) We were able to read the blocks while waiting for the neutron probe readings. Both gypsum blocks and the neutron probe were read in less than five minutes at each site.

3) It is apparent that the blocks react differently depending on soil type. A certain insensitivity at moist conditions indicates that for coarse soils the blocks must be used in sets. In coarse soils there is no indication of drying until very near the irrigation point. In hot weather, one block may drop from a wet to a dry reading in one or two days. However, if the blocks are used in sets on the top several feet of the root zone, the upper blocks will show a reading before the lower blocks and give ample warning when an irrigation is about to be needed.

4) We used a meter with a single scale. Subsequent devices have multiple scales that may give more sensitivity in more moist conditions.

5) Salinity conditions were encountered. Indications of this condition can be identified from the readings. This would not seem to reduce the utility of the device.

6) While each soil would need a calibration to convert meter readings into soil moisture, as experience is gained with the particular field the readings can be used directly to determine water-penetration depth and the proper time to irrigate. Determining how much to irrigate would require a calibration, but with first-hand experience and a knowledge of the amounts applied this information might also be determined from field use.

I hope these comments will be of value to you.

Gerald A. Robb
Water Conservation & Management Specialist

G
Report on 1987 Field Demonstrations of the Soil-Moisture Method Sponsored by the Yolo County Resource Conservation District

The following report summary (slightly shortened) and the table of findings, were sent to INFORM by John Tiedeman, SCS Agricultural Engineer, who conducted field demonstrations of the soil-moisture method for the Yolo County Resource Conservation District in 1987:

In 1987 the Yolo County Resource Conservation District sponsored an irrigation water-management program to demonstrate a low-cost, practical approach to scheduling and managing surface-irrigation on alfalfa, using gypsum blocks. The demonstration was funded by INFORM, a non-profit organization promoting research for environmental protection.

Ten out of 14 fields in the study were located in Yolo County, three in Solano County, and one in Colusa County. Support staff included technicians from each of the three SCS/RCD county offices plus staff from the

G: Report on 1987 Field Demonstrations

Yolo County Flood Control and Water Conservation District.

The field demonstrations explored a range of representative surface irrigation problems and solutions and introduced a core of technical field staff in the three-county study area to the gypsum-block method.

The tests revealed that two opportunities for irrigation improvement are by far the most common: (1) early season irrigations are applied when they are not needed, and (2) poor water distribution uniformity results from shutting water off completely before it advances to the end of the field. The gypsum-block soil-moisture data provide the necessary information for addressing both of the above situations.

Out of 14 alfalfa fields in the gypsum-block field demonstrations, the following results were observed: 1) three fields had water distribution uniformity problems (from TOP to END of field), which after correction produced 16-20% yield increases; 2) five fields showed water savings ranging from 20 to 44%; 3) six fields were doing as well as possible due to soil or other limitations. The following table summarizes the results by individual field.

Appendices

RESULTS OF 1987 GYPSUM BLOCK FIELD DEMONSTRATIONS[1]

	Operator	Distribution Problem Corrected[2]	Yields Increased	Water Saved	Other
1	Schaad (south) Yolo County	x	20%		
2	Schaad (north) Yolo County	x	20%		
3	Detling Yolo County			x	
4	Peterson Yolo County				Permeability study indicated.
5	Tadlock Yolo County			30%	
6	Hilleby Yolo County	x	16%		
7	Gnoss Yolo County		x	x	
8	Chamberlain (south) Yolo County				Drainage problem corrected.
9	Chamberlain (north) Yolo County				Doing as well as possible.
10	Gyorke Yolo County				Doing as well as possible.
11	Braun (south) Solano County				Doing as well as possible. [3]
12	Braun (north) Solano County			44%	
13	Papin Solano County				Doing as well as possible.
14	Ash Brothers Colusa County			20%	

1. Results were based only on comparison of test and control areas, not entire fields.
2. Some additional water usage incurred with cutback irrigation for distribution problems.
3. One irrigation/cutting mid-season was infeasible due to extended cutting cycle (35-40 days vs. normal 28-30 days).

About the Authors

Gail Richardson, Ph.D.

Since 1982, Dr. Gail Richardson has developed and directed INFORM's field study of the soil-moisture method of managing surface irrigation. In 1985, she wrote INFORM's report, *Saving Water From the Ground Up: Pilot Tests of Irrigation Scheduling on Four California Fields*. Dr. Richardson has served as INFORM's Director of Development, Director of Research and Associate Director. She is presently a Senior Research Consultant to INFORM.

Dr. Richardson received a Ph.D. in Political Science from the University of Wisconsin; a Masters in Law and Diplomacy from the Fletcher School; and a Bachelor of Arts degree in English literature from Cornell University. She has taught political theory and comparative politics at the City College of New York and the American University of Rome, Italy.

Peter Mueller-Beilschmidt, P.E.

Peter Mueller-Beilschmidt, an independent engineer and consultant in California, has 30 years of experience in engineering irrigation systems and improving irriga-

About the Authors

tion management. From 1956 to 1969, Mr. Mueller-Beilschmidt developed and managed the irrigation contracting division of Vicco, Inc. In this capacity he designed and installed California's first hose-pulled sprinkler system for citrus orchards, and the state's first solid-set, over-the-vine sprinkler system for vineyards.

Mr. Mueller-Beilschmidt received his Master of Science degree in agricultural engineering from the University of Giessen in West Germany, and his Bachelor of Science in agriculture from the University of Jena in East Germany.

INFORM's Board of Directors

Kenneth F. Mountcastle, Jr.
Chairman
Senior Vice President
Dean Witter Reynolds,
  Incorporated

Linda Stamato
Vice Chairman
Associate Director
Center for Negotiation and
  Conflict Resolution
Rutgers University

James B. Adler
President
Adler & Adler Publishing
  Company

Michael J. Feeley
President &
Chief Executive Officer
Feeley & Willcox

Barbara D. Fiorito
Vice President
Discount Corporation
  of New York Advisers

Jane R. Fitzgibbon
Senior Vice President
Head of Research
Ogilvy & Mather Advertising

C. Howard Hardesty, Jr.
Partner
Andrews & Kurth

Timothy L. Hogen
President
T.L. Hogen Associates

Lawrence S. Huntington
Chairman of the Board
Fiduciary Trust Company
  of New York

Sue W. Kelly
Lecturer
Graduate Program
Sarah Lawrence College

Martin Krasney

Jay T. Last

Charles A. Moran
President
Government Securities
  Clearing Corporation

Susan Reichman

Frank T. Thoelen
Partner
Arthur Andersen
  & Company

Grant P. Thompson
Executive Director
League of Women
  Voters

Edward H. Tuck
President
The French-American
  Foundation

Joanna D. Underwood
Executive Director
INFORM

Frank A. Weil
Chairman &
Chief Executive Officer
Abacus & Associates

Anthony Wolff
Writer and Photographer

Recent INFORM Publications

To order any of the following publications please indicate below the quantity of each. Substantial discounts for 5 or more—please write for information. All orders must be prepaid. Please fill out both sides of this form.

SOLID WASTE MANAGEMENT

___ *Garbage Management in Japan: Leading the Way*

by Allen Hershkowitz, Ph.D. and Eugene Salerni, Ph.d. (1987, 130 pp) $15.00

___ *Garbage Burning: Lessons from Europe: Consensus and Controversy in Four European States*

by Allen Hershkowitz, Ph.D. (1986, 53pp) $9.50

___ *Garbage: Practices, Problems & Remedies*

by Joanna D. Underwood, Allen Hershkowitz, Ph.D. and Maarten de Kadt, Ph.D. (1988, 25pp) $3.50

TOXIC WASTES

___ *Tracing a River's Toxic Pollution: A Case Study of the Hudson - Phase II*

by Steven O. Rohmann, Ph.D. and Nancy Lilienthal (1987, 209pp) $19.95

___ *Tracing a River's Toxic Pollution: A Case Study of the Hudson*

by Steven O. Rohmann, Ph.D. (1985, 150pp) $12.00

Set $25.00

___ *Promoting Hazardous Waste Reduction: Six Steps States Can Take*

by Warren R. Muir, Ph.D. and Joanna D. Underwood (1987, 21pp) $3.50

___ *Cutting Chemical Wastes: What 29 Organic Chemical Plants are Doing to Reduce Hazardous Wastes*

by David J. Sarokin, Warren R. Muir, Ph.D., Catherine G. Miller, Ph.D. and Sebastian R. Sperber (1986, 535pp) $47.50

___ *Tracking Toxic Wastes in New Jersey (and in California and in Ohio).* Three separate Guides to government information sources

by Catherine G. Miller, Ph.D. and Laurence M. Naviasky (1986) $15.00 each

OTHER

___ *Controlling Acid Rain: A New View of Responsibility*

by James S. Cannon (1987, 55pp) $9.95

___ *A Directory of Independent Workers' Clinics*

by Laurence M. Naviasky (1986, 36pp) $4.95

Subtotal: $ _____

Postage and handling $2.50

Total $ _____

Forthcoming Publications

How-To Guide for Farmers on the Soil-Moisture Method of Irrigation Management

Study of 15 U.S. Garbage Burning Plants

Update on Waste Reduction Activities at the 29 Chemical Plants Studied in Cutting Chemical Wastes

Changes in Environmental Performance and Planning at Nine Florida Subdivisions Since the Mid-1970's

How Methane and Methanol use as Motor Vehicle Fuels Might Reduce Urban Air Pollution

Prices and approximate dates of release of forthcoming publications are available upon request.

Enclosed is my check for $ _____

For orders over $10.00 only: ☐ VISA ☐ MASTER CARD ☐ AMEX
Card # _____ Expiration Date _____

Signature _____

Name _____ Title _____
Company/Affiliation _____
Address _____
City _____ State _____ Zip _____

☐ I would like to become a member of INFORM. A $25.00 membership fee entitles me to a one year subscription to INFORM's bi-monthly newsletter, *INFORM Reports*.

Please make checks payable to INFORM. Mail to INFORM, 381 Park Avenue South, New York, NY 10016 (212)689-4040